Dynamic Approaches to Global Economic Challenges

Birgit Bednar-Friedl • Jörn Kleinert
Editors

Dynamic Approaches to Global Economic Challenges

Festschrift in Honor of Karl Farmer

Springer

Editors
Birgit Bednar-Friedl
Department of Economics & Wegener
Center for Climate and Global Change
University of Graz
Graz, Austria

Jörn Kleinert
Department of Economics
University of Graz
Graz, Austria

ISBN 978-3-319-23323-9 ISBN 978-3-319-23324-6 (eBook)
DOI 10.1007/978-3-319-23324-6

Library of Congress Control Number: 2015951447

Springer Cham Heidelberg New York Dordrecht London

Printed on acid-free paper

Springer International Publishing AG Switzerland is part of Springer Science+Business Media (www.springer.com)

Contents

Contributors

Birgit Bednar-Friedl Department of Economics, University of Graz, Graz, Austria;
Wegener Center for Climate and Global Change, University of Graz, Graz, Austria

Dmitri Blueschke Department of Economics, Alpen-Adria-Universität Klagenfurt, Klagenfurt, Austria

Anna Boisits Center for Accounting Research, University of Graz, Graz, Austria

Fritz Breuss Vienna University of Economics and Business, Institute for International Economics, Vienna, Austria;
Austrian Institute of Economic Research, Macroeconomics and European Economic Policy, Arsenal, Vienna, Austria

Pasquale Commendatore Department of Law, University of Naples Federico II, Naples, Italy

Jörn Kleinert Department of Economics, University of Graz, Graz, Austria

Ingrid Kubin Department of Economics, WU Vienna University of Economics and Business, Vienna, Austria

Wolf Rauch Department of Information Science and Information Systems, University of Graz, Graz, Austria

Matthias Schelnast Department of Economics, University of Graz, Graz, Austria

Thomas Schinko Wegener Center for Climate and Global Change, University of Graz, Graz, Austria;
International Institute for Applied Systems Analysis (IIASA), Laxenburg, Austria

Stefan P. Schleicher Wegener Center for Climate and Global Change, University of Graz, Graz, Austria

Karl W. Steininger Department of Economics, University of Graz, Graz, Austria;
Wegener Center for Climate and Global Change, University of Graz, Graz, Austria

Iryna Sushko National Academy of Science of Ukraine, Institute of Mathematics, Kyiv, Ukraine;
Kyiw School of Economics, Kyiw, Ukraine

Ronald Wendner Department of Economics, University of Graz, Graz, Austria

Jacopo Zotti Fondazione Eni Enrico Mattei, Venice, Italy;
Department of Political and Social Sciences, University of Trieste, Trieste, Italy

Introduction

Birgit Bednar-Friedl and Jörn Kleinert

The ten articles in the book to be presented to Karl Farmer's 65th birthday cover a wide range of topics in three areas of economic research: international economics, structural change, and environmental and resource economics. In these fields, Karl made his footprints particularly in the city of Graz, province of Styria, Austria where he has worked for 35 years, but also beyond Styria's capital. Karl Farmer, the trained mathematician and economist, has shaped generations of students in their approach towards economic theory which demands exactness and precision in theoretical arguments. His success as a teacher is evident in this Festschrift by contributions of a selection of his former students of which some become colleagues later. He was not only extremely influential for generations of economics and business students at the University of Graz, but he has served as a visiting professor at the University of Natural Resources and Life Sciences, Vienna, Austria, and at the University of Cluj, Romania, for many years. His dedication to teaching is also demonstrated in his textbooks on international macroeconomics and on resource economics (Farmer and Bednar-Friedl, 2010; Farmer and Schelnast, 2013; Farmer and Vlk, 2011; Farmer and Wendner, 1999).

Karl is at home in general equilibrium theory, particularly in intertemporal applications. He has used this tool in many studies on topics in one of the three fields in the volume. In recent years, he has been particularly concerned with debt limits in open economies. The crisis at the European periphery motivated and affected his research in many respects. We are therefore sure that the first part in the book not only meets his research interest but also refreshes and intensifies discussions on the European integration process. He continuously was also interested in understanding the process of technological change and growth. Building on his earlier work in resource economics and sustainability, during recent years Karl's work increasingly

B. Bednar-Friedl • J. Kleinert (✉)
Department of Economics, University of Graz, Universitätsstraße 15, 8010 Graz, Austria
e-mail: birgit.friedl@uni-graz.at; joern.kleinert@uni-graz.at

B. Bednar-Friedl, J. Kleinert (eds.), *Dynamic Approaches to Global Economic Challenges*, DOI 10.1007/978-3-319-23324-6_1

centered on the problem of climate change. Here, his interest was primarily in exploring the international trade implications of climate policy. The second part of this book therefore provides a collection of articles which on the one hand discuss the issue of technological and structural change, and on the other climate change policy.

The contribution “A Prototype Model of European Integration: The Case of Austria” by Fritz Breuss (2015) opens the discussion on European integration. The paper offers an ex-post evaluation of four integration steps of Austria in the European Union (EU): (i) the opening-up of Eastern Europe after the fall of the Iron Curtain in 1989, (ii) the Austrian EU accession in 1995, (iii) Austria’s participation in the Economic and Monetary Union (EMU) from the beginning in 1999, and (iv) the EU Eastern enlargement in 2004 and 2007. Fritz Breuss proposes a small integration model for Austria which can be used to assess qualitative effects. The model comprises integration effects from goods and factor markets, from changes in competition, from effects of adjusted institutions, and the changing global environment. In times of the Eurozone crisis and discussion of Grexit (Greek exit), Austrian economist Breuss (2015) reminds us of the large positive effects that Austria’s integration in the EU has had on the Austrian economy. He estimates an overall integration gain of about 0.9 % points of additional annual growth through EU integration, a falling unemployment rate, and lower inflation. These are not only large effects but also continuing ones. For Austria, the integration in the EU has the effect of a long-lasting boom for two decades.

The Italian economy had its integration boom in the 1960s. In the last integration steps the Italian economy has struggled more with the new competitors and rules of competition than it has used new opportunities. The reason for this disappointing performance might be found in the oligopolistic market structure in the non-tradable goods sector, as proposed by Jacopo Zotti (2015). He presents an application of the theory of the second best, when explaining the problems of the Italian tradable good sector by the interrelation of economic integration and oligopolistic market structures. Economic integration is driven by the outsourcing of components in the competitive tradable goods sector. From a welfare perspective, integration increases too much, i.e., outsourcing goes too far, in response to a fall in barriers to trade and internationalization if the non-tradable sector is oligopolistic. The tradable goods sector experiences falling prices and increasing output while the oligopolistic sector strategically reduces output to increase prices. The contribution highlights the important role of market structure analysis and competition policy in assessing welfare effects of integration processes, regardless whether they occur at global or at regional level.

As in Zotti’s (2015) contribution with respect to outsourcing, changes in the location of production are at the center of interest in the paper by Commendatore, Kubin and Sushko (Commendatore et al., 2015). They model the integration process in the EU in a three-country model with two countries reducing barriers to trade, and relocation of firms respectively while the third country is apart from the integration process. Firms have monopolistic power in a niche of the market. They are defined by their entrepreneurs who are free to migrate between the countries. Thus, firms are footloose. Migration is costly however and costs differ between the countries and change in the integration process. Deeper integration is characterized by lower migration and lower trade costs in the model. Specialization, trade creation and diversion effects result from the integration process. In addition, agglomeration may arise within the union which splits the union in a core-periphery pattern.

While dynamics play a role in the contribution by Commendatore et al. (2015) they are not as pronounced as in the work by Neck and Blueschke (2015). Neck and Blueschke (2015) study the interaction of national fiscal policy and supra-national monetary policy in a dynamic tracking game framework. They show that a cooperative solution for the strategic game outperforms the non-cooperative Nash solution. They then introduce a sequence of negative demand side shocks to mimic the recent crisis in order to analyze possible policy reactions to this challenge. A countercyclical and mildly expansionary monetary policy turns out to be the optimal policy response. The challenge is to avoid excessive public debts which threaten the solvency of the union.

Such debt ceilings are in the center of the contribution by Boisits and Schelnast (2015). They study the linkage between maximum sustainable government debt limits and other economic variables, such as saving rates and public expenditures in order to analyze the impact of these variables on the maximum sustainable government debt level. They show that higher saving rates induce an increase in debt limits and confirm that domestic debt limits are negatively related to foreign debt ratios. Government expenditures affect the maximum sustainable debt ratio negatively. These results that build on early work by Karl Farmer (Farmer, 2006; Farmer and Schelnast, 2013; Farmer and Zotti, 2010) are very important for further discussions on a common fiscal policy in the EMU.

The second part of this book starts with a stochastic dynamic general equilibrium model with heterogeneous households. With this model, Kleinert and Wendner (2015) analyze the effects of positional preferences on the interaction between the distribution of wealth and endogenous economic growth. Households exhibit positional preferences, i.e., they derive utility not only from their own consumption level, they are also concerned about their wealth rank, i.e., their wealth–relative position in the distribution of wealth in society. Contrary to adopting ad hoc assumptions regarding the specification of the wealth ranks, Kleinert and Wendner (2015) demonstrate that the distribution of wealth ranks necessarily follows a power law. Under this power law, individual wealth growth rates differ, though, the aggregate distribution converges to a stationary distribution in the balanced growth path. Differences in individual growth rates are due to differences in the return to households' assets (e.g., differences in the return to one's human capital—considered as part of the total capital). Moreover, the endogenous growth rate is shown to strongly depend on the *power law exponent*, which is itself an endogenous variable that is determined by the standard deviation of the individual growth rates and the minimum level of the wealth distribution.

This model based explanation of economic growth is then followed by a review by Rauch (2015) on technological change in communication and information technologies and how these digital technologies have and eventually will affect research. Wolf Rauch, an information scientist, argues that we are currently in a phase of transition similar to the transition from the oral to the written word in ancient times. He addresses both the threats, such as the problem of data storage, and the opportunities, such as the availability of big data which allow for new statistical approaches which are no longer hampered by small sample size. He finally argues that analysis of big data might change the way we conduct research, in that causality as central scientific method might become obsolete.

Technological change is also the topic of the contribution by Stefan Schleicher (2015). He discusses how traditional macroeconomic models fail to incorporate both the stock and flow properties of capital stocks, resource stocks and materials. He argues that an alternative approach which considers both stocks and flows provides interesting policy conclusions.

The last two articles in this contribution focus on the problem of climate change. Bednar-Friedl (2015) discusses how sequential policy setting affects the choices of first and second movers in climate policy. While this question has been addressed in the game–theoretic literature before, she uses a two country overlapping generations model. The advantage of this latter approach is that countries do not only differ in marginal damage and marginal abatement costs, as is the case with game–theoretic models due to their partial equilibrium setup, but that countries also differ in economic variables such as net foreign asset positions, saving rates and energy intensity of production. She illustrates how these differences translate into different abatement levels, both among industrialized countries like the USA and the European Union, but also between industrialized and developing countries.

Steininger and Schinko (2015) discuss a related topic: the question of how carbon dioxide emissions have evolved both domestically and abroad. While regions such as the EU have been successful in reducing emissions within their own territory, the emissions triggered by imports to the EU have risen steadily over the last few decades. Steininger and Schinko (2015) therefore discuss how alternatives to the current accounting system of the United Nations Framework Convention on Climate Change (UNFCCC), which only takes account of emissions within the own territory but not of emissions caused abroad, could be changed to counter this increasing difference between domestic and foreign emissions.

Finally, on a personal note we would like to express our deepest gratitude for having had the chance to work with Karl. As head of the Department of Economics and Vice Dean of Studies, he has steered us through rocky waters for more than a decade. He always had an open ear for everyone, be it colleagues or students. Being relieved from the duties of everyday work, we very much hope that he can now spend more time with his wife, children and grandchildren and also continue to work on his research he always has cared so much about.

References

Bednar-Friedl B (2015) Is there a first-mover advantage in international climate policy? In: Bednar-Friedl B, Kleinert J (eds) Dynamic approaches to global economic challenges. A Festschrift in Honor of Karl Farmer. Springer, Berlin, Heidelberg, pp 155–170

Boisits A, Schelnast M (2015) Determinants of maximum sustainable government debt. In: Bednar-Friedl B, Kleinert J (eds) Dynamic approaches to global economic challenges. A Festschrift in Honor of Karl Farmer. Springer, Berlin, Heidelberg, pp 75–92

Breuss F (2015) Trade agreements and regional disparities. In: Bednar-Friedl B, Kleinert J (eds) Dynamic approaches to global economic challenges. A Festschrift in Honor of Karl Farmer. Springer, Berlin, Heidelberg, pp 9–30

Commendatore P, Kubin I, Sushko I (2015) Trade agreements and regional disparities. In: Bednar-Friedl B, Kleinert J (eds) Dynamic approaches to global economic challenges. A Festschrift in Honor of Karl Farmer. Springer, Berlin, Heidelberg, pp 31–52

Farmer K (2006) Reducing public debt under dynamic efficiency: transitional dynamics in diamond's olg model. Atl Econ J 34(2):195–208

Farmer K, Bednar-Friedl B (2010) Intertemporal resource economics. An introduction to the overlapping generations approach. Springer, Heidelberg

Farmer K, Schelnast M (2013) Public debt reduction in advanced countries and its impact on emerging countries. Int Adv Econ Res 19(2):167–188

Farmer K, Vlk T (2011) Internationale Ökonomik. Eine Einführung in die Theorie und Empirie der Weltwirtschaft, 4th edn. LIT Verlag, Wien

Farmer K, Wendner R (1999) Wachstum und Außenhandel. Eine Einführung in die Gleichgewichtstheorie der Wachstums- und Außenhandelsdynamik, 2nd edn. Springer, Heidelberg

Farmer K, Zotti J (2010) Sustainable government debt in a two-good, two-country overlapping generations model. Int Rev Econ 57(3):289–316

Kleinert J, Wendner R (2015) Status, wealth distribution, and endogenous economic growth. In: Bednar-Friedl B, Kleinert J (eds) Dynamic approaches to global economic challenges. A Festschrift in Honor of Karl Farmer. Springer, Berlin, Heidelberg, pp 117–129

Neck R, Blueschke D (2015) Strategic macroeconomic policies in a monetary union. In: Bednar-Friedl B, Kleinert J (eds) Dynamic approaches to global economic challenges. A Festschrift in Honor of Karl Farmer. Springer, Berlin, Heidelberg, pp 53–73

Rauch W (2015) Technological change in information and communication: Consequences for science. In: Bednar-Friedl B, Kleinert J (eds) Dynamic approaches to global economic challenges. A Festschrift in Honor of Karl Farmer. Springer, Berlin, Heidelberg, pp 131–140

Schleicher SP (2015) The economics of climate change or the climate change of economics. In: Bednar-Friedl B, Kleinert J (eds) Dynamic approaches to global economic challenges. A Festschrift in Honor of Karl Farmer. Springer, Berlin, Heidelberg, pp 141–154

Steininger KW, Schinko T (2015) Environmental policy in an open economy: Refocusing climate policy to address international trade spillover. In: Bednar-Friedl B, Kleinert J (eds) Dynamic approaches to global economic challenges. A Festschrift in Honor of Karl Farmer. Springer, Berlin, Heidelberg, pp 171–190

Zotti J (2015) The long italian stagnation and the welfare effects of outsourcing. In: Bednar-Friedl B, Kleinert J (eds) Dynamic approaches to global economic challenges. A Festschrift in Honor of Karl Farmer. Springer, Berlin, Heidelberg, pp 93–113

Part I
Real, Monetary and Fiscal Integration in the European Union

A Prototype Model of European Integration: The Case of Austria

Fritz Breuss

1 Introduction

As an EU member state since 1995, Austria has taken part in all subsequent European integration steps: the *deepening* of EU integration via the single market and Economic and Monetary Union (EMU) with the introduction of the euro, and the *enlargement* of the EU. Since the accession of Croatia in 2013, the EU encompasses 28 member states, 19 of which have introduced the euro.

Around the years 2014 and 2015, Austria has had and will have numerous anniversaries to celebrate in connection with European integration: 25 years since the fall of the Iron Curtain and hence the opening up of Eastern Europe in 1989; 20 years of EU membership (1995); 15 years of Austria's EMU membership (1999), and 10 years of EU enlargement (starting in 2004).

In this contribution, an integration model for Austria is developed to estimate empirically the integration effects since 1989. It is an econometrically estimated macro model, capturing the main features of European integration since the opening up of Eastern Europe in 1989, namely encompassing the effects of EU accession in 1995 and the participation in EMU since 1999, as well as the effects of EU enlargement since 2004. This small integration model for Austria could also serve as a prototype for other member states of the EU.

F. Breuss (✉)
Vienna University of Economics and Business, Institute for International Economics, Welthandelsplatz 1, 1020 Vienna, Austria

Austrian Institute of Economic Research, Macroeconomics and European Economic Policy, Arsenal, Object 20, 1030 Vienna, Austria
e-mail: Fritz.Breuss@wu.ac.at; Fritz.Breuss@wifo.ac.at

B. Bednar-Friedl, J. Kleinert (eds.), *Dynamic Approaches to Global Economic Challenges*, DOI 10.1007/978-3-319-23324-6_2

2 Austria's Integration into Europe

At the beginning of each integration step, several studies were undertaken in the EU and also in Austria to estimate ex ante the possible integration effects (for an overview of such studies, see Breuss 2012, p. 43). In the Austrian studies by the Austrian Institute of Economic Research (WIFO), simulations were carried out by means of the—at the time—actual version of the WIFO macroeconomic model.[1] This study employs a macro integration model to evaluate ex post the integration effects Austria has realized since the opening up of Eastern Europe in 1989, and in particular the economic impact of EU accession in 1995.

Over the last decades, European integration has systematically progressed from a customs union (completed in 1968) towards the single market, EMU, and major EU enlargements. Integration theory was either applied ahead of the actual implementation, as in the case of the EC's custom union (see Viner 1950), or had to catch up with European integration as it progressed towards the single market and the EMU projects (see Baldwin and Venables 1995; Breuss 2003a; or for a survey, see Jovanovic 2011).

Austria has taken part in all integration steps since the opening up of Eastern Europe in 1989, gaining EU membership in 1995 and EMU membership in 1999, and participating as an EU member in the EU enlargements since 2004. An overview of the possible theoretical integration effects[2] in the case of Austria's EU integration is illustrated in Fig. 1.

Economies of scale (EOS) played an important role at the stage of creating the single market, as did competition effects via the harmonization of competition rules concerning a common legal base. The liberalization of certain sectors and privatization were also part of the single market program. Other effects derived from the implementation of the Common Agricultural Policy (CAP), the common foreign trade policy (as a consequence of the customs union and the dismantling of border controls), and the harmonization of other policies, such as regional or structural policies. There is also the EU budget, which finances the different policy areas with a view to ensuring solidarity between member countries, implying a redistribution of funds from 'rich' EU members (net contributors) to 'poor' ones (net recipients). Overall, the single market is supposed to boost intra-EU trade and, via gains in efficiency and productivity, lead to stronger economic growth.

[1] In a comprehensive ex ante study Breuss et al. (1994) estimated the potential macroeconomic and sectoral effects of Austria's EU accession with the WIFO macro cum input–output model. Keuschnigg and Kohler (1996) estimated, also ex ante, the possible Austrian integration effects of EU accession with a single-country dynamic general equilibrium model (sectoral and macroeconomic results).

[2] A detailed overview of integration theories applicable for the several steps of European integration can be found in Badinger and Breuss (2011) and in Breuss (2014).

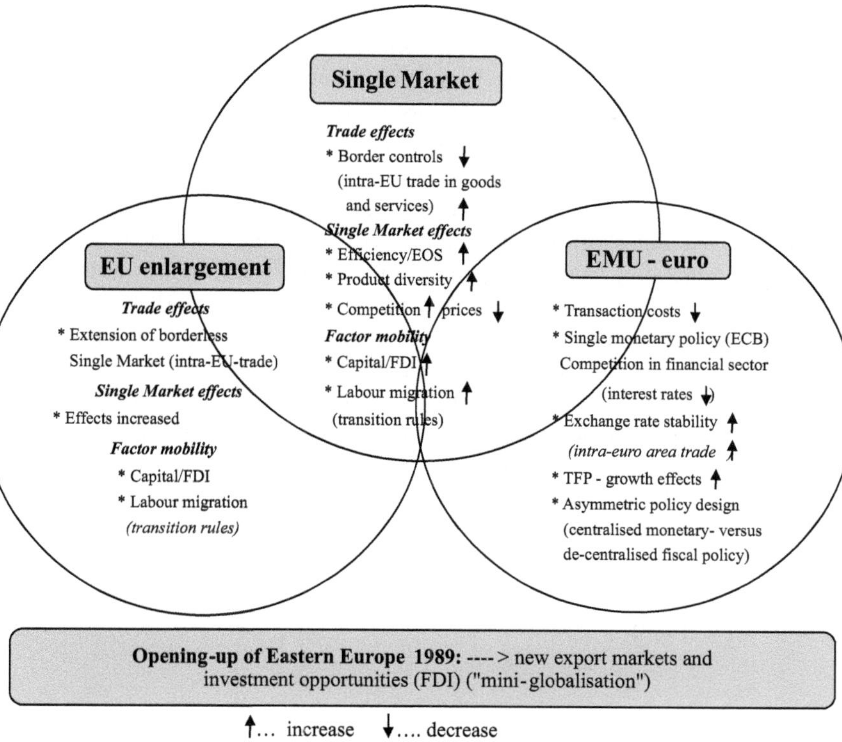

EOS . . . economies of scale; FDI . . . foreign direct investments; TFP . . . total factor productivity.

Fig. 1 Effects of Austria's EU integration. (Overview of the theoretical integration effects)

Across the large number of existing integration studies, single market effects are estimated using different methods and approaches: macroeconomic models and/or microeconomic models, both for individual countries (country studies with single-country models) and/or for several countries (multi-country models). Among the model approaches there are macro models or general equilibrium models. Within the modern theory of endogenous growth, there are special derivations for the growth effects of integration (see Breuss 2003a).

A step that is more complicated is the assessment of the integration effects originating from the EU's Economic and Monetary Union (EMU) and the introduction of the euro as a common currency. In this respect, theory is virtually entering uncharted waters. A relatively well-developed theory is that of 'optimal currency areas' (OCAs), exploring which countries would be in a sustainable position to share a common currency. Early studies arrived at the conclusion that in Europe only a small number of OCAs would be able to survive (see Breuss 2006). As the current euro-area crisis painfully demonstrates, the euro project was driven by political considerations rather than being founded on the basis of sound economic criteria.

Shortly before the fourth round of EU enlargement to include Austria, Finland, and Sweden in 1995, the EU—confronted with historical events such as the collapse of communism and the fall of the Iron Curtain—was virtually forced to integrate the former Soviet satellite countries of Eastern Europe. Those countries were then gradually integrated into the EU single market, first by Europe agreements and later by formal EU accessions, starting in 2004.

The effects of the three integration steps, the single market, the EMU, and EU enlargement, overlap, as illustrated in Fig. 1. Austria, then still a member of EFTA, was already benefiting from the opening up of Eastern Europe towards the West in 1989. This event suddenly facilitated access to Eastern markets that hitherto had been severely constrained by the 'Iron Curtain,' offering new opportunities for export and foreign direct investment (FDI). Since the opening up of Eastern Europe, Austria has to a greater extent than before taken part in globalization ('mini-globalization') as it has moved from a marginal position into the center of Europe. Austria's accession to the EU in 1995 and to the EMU in 1999 augmented the benefits already obtained through the opening up of new markets via the 1989 revolution. These effects were reinforced again by the EU enlargement rounds of 2004 and 2007. Austria's ever deeper integration into the EU has, via the operation of the manifold integration effects, in almost all cases led to higher economic growth and greater prosperity.

3 An Integration Model for Austria

To evaluate quantitatively ex post the integration effects Austria has realized in the past, an integration model has been designed. The integration effects derived in this way represent the deviations of actual economic developments in Austria from the hypothetical path that the economy would have followed had Austria remained apart from all integration moves since 1989. The integration model for Austria applied here is a small macro model econometrically estimated with EViews 7.0 for a dataset over the period 1960–2015. The detailed set of equations can be found in Appendix 1. This integration model for Austria could also be used as a prototype for other EU member states to evaluate their integration effects.[3]

[3] To estimate ex post the integration effects of Austria's EU membership, on earlier occasions a similar small country macro model approach has been applied (Breuss 2010a, 2013c). In the case of a comparison of the integration performance of Austria, Finland, and Sweden in the EU (Breuss 2003b), and for the evaluation of the EU accession of Bulgaria and Romania (Breuss 2010b), small macro-integration models of a similar type as the present integration model were estimated to simulate the specific integration features of these countries.

Table 1 Model inputs for simulating integration effects for Austria

Scenarios		Integration effects	Model inputs
1	Opening up 1989	Trade and FDI Net migration	Regime change T+FDI Migration 1989–1993
2	EU member 1995	Increased price competition TFP-stimulating R&D Trade and FDI EU net budget position	Mark-up decreasing since 1995 Regime change R&D Regime change T+FDI Av. −0.25 % GDP since 1995 Migration 1995–2015
3	EMU member 1999	More competitiveness trade and FDI TFP-stimulating R&D	No appreciation since 1999 Regime change T+FDI Regime change R&D
4	EU enlargement 2004/07	Trade and FDI Net migration	Regime change T+FDI Migration 2004–2015

'Regime change T+FDI' = regime change dummy variable for trade and FDI; 'Regime change R&D' = regime change dummy variable for research and development (R&D); TFP = total factor productivity. For detailed data inputs, see Appendix 2

3.1 *Four Integration Steps Since 1989*

We evaluate the integration effects of Austria's European integration with reference to the major variables of the macro model, in particular to the impact on real GDP. Real GDP per capita is our final 'welfare' measure. The four steps of Austria's deep integration into Europe since 1989 is evaluated in scenarios (see Table 1).

3.1.1 Opening Up of Eastern Europe in 1989

The opening up of Eastern Europe in 1989 increased the potential of Austria's markets for direct trade and FDI, and implied a net inflow of migrants. Scenario 1 therefore takes into account two effects:

- *Trade and FDI effects*: To capture the trade and FDI effects, we introduce a 'regime change' variable[4] ('Regime change T+FDI' or the dummy 'D_1989_2015' in the trade and FDI equations in the integration model of Appendix 1), which takes the value of 1 until 1988, then increases by 0.1 in each following integration step; thus, in 1989 it increases to 1.1, remains at this level until 1994, and jumps to 1.2 in 1995, in 1999 to 1.3, in 2004 to 1.4, and in 2007 to 1.5, remaining at this level until 2015. In the simulations of the

[4]The literature treats 'regime changes' in the context of 'regime switching models' using Markov chain econometrics (e.g., see Hamilton 2008). Generally, many economic time series occasionally exhibit dramatic breaks in their behaviour, associated with events such as financial or other crises. In our case, the breaks occurred due to four integration shocks (1989, 1995, 1999, and 2004/07) of European integration.

'opening-up' scenario, the regime change dummy was reduced to 1 from 1989 until EU enlargement began (see Appendix 2). This regime change dummy can be interpreted as a 'smart dummy' (capturing price and non-price effects of trade liberalization vis-à-vis the Central and Eastern European Countries, CEECs) and is included in the estimations of the equations for real exports and imports, for FDI exports and imports. In line with the insights of the 'New' New Trade Theory (see the application in the TTIP evaluation by Felbermayr et al. 2013), in our model more trade engagement translates indirectly via the R&D equation into an increase in total factor productivity (TFP) and hence has an accumulation or growth effect, leading to higher real GDP.

- *Net migration*: Besides the trade and FDI effects, the opening up of Eastern Europe in 1989 also had net migration effects. As can be seen from Appendix 2, the biggest net inflow of migrants occurred shortly after the collapse of the former Yugoslavia in the early 1990s. In the integration model, net migration inflows enter exogenously via the unemployment equation into labor supply. Migration also affects the definition of GDP per capita via the variable population (see Appendix 1).

3.1.2 EU Membership in 1995

A new EU member must take over the acquis communautaire (Community acquis) of the single market project. This implies communitization, i.e., the transfer of competencies, from former national responsibility to EU competence in many economic policy areas: the CAP, the Common Commercial Policy (CCP) by entering into the EU customs union, the common competition policy, and a common regional/structural policy, and many other areas in which economic policy is harmonized at the EU level.

In scenario 2, Austria's EU membership is captured by five inputs:

- *Greater price competition*: Entering into the single market increases price competitiveness, which is captured by reducing the mark-up on unit labor costs.[5] We assume that the mark-up in the case of Austria's EU membership increased strongly in the beginning and tapered off later. In the simulations, the dummy variable for price mark-up was reduced in value from 1.3 pre-accession to 1.2 in 1995, to 1.1 in 1996, and to 1.0 in 1997 (see Appendix 2). The main result is that consumer prices decline, but the real GDP effects are negligible.
- *TFP-stimulating R&D expenditures*: EU membership has improved the opportunities for Austrian research institutions (universities and non-university institutions and firms) to participate fully in EU research programs (Framework

[5] Badinger and Breuss (2005) analyzed the sectoral change of mark-up pricing after EU accession in Austria. The results were mixed. Some sectors had pronounced mark-up reductions (mining and quarrying, wholesale and retail trade, financial services and real estate), whereas in other sectors no notable mark-up change was found.

Programmes). This resulted in a break in the trend of R&D expenditures in % of GDP. After EU accession, the R&D trend was much steeper than in the pre-EU period. These additional R&D opportunities are captured by another 'smart dummy,' namely the variable 'Regime change of R&D' (or the dummy 'D_1995_2015' in the R&D equation in the integration model; see Appendix 1). Due to participation in the EU's research programs, the R&D dummy jumps in 1995 from 1 to 1.1 (see Appendix 2). In our model context, more R&D stimulates TFP and hence real GDP growth.

- *Trade and FDI*: A country entering the EU and hence the single market must also enter into the EU customs union with its common external tariff (CET). In the case of Austria, this implied a reduction in the average tariff rate from 10.5 % to the CET level of 5.7 % before the cut in the context of the Uruguay Round in 1995. Besides the minor reduction in import tariffs, the major reduction concerned the abolition of border controls, resulting in cost savings for firms engaged in foreign trade. All price and non-price (NTBs) changes in connection with EU accession should be captured with our 'smart dummy' variable 'Regime change T+FDI.' In the simulations, the T+FDI dummy was increased from 1.1 to 1.2 in 1995 (see Appendix 2). Participation in the EU's single market of course improves the opportunities to expand foreign trade. But this is not a one-sided affair. The opening up of borders (abolition of border controls) drives competitive importers into the market of the newcomer. On balance, Austria gained from full participation in the CAP, but overall the trade balance vis-a-vis the EU deteriorated following 1995. In addition to trade, bilateral FDI flows also increased after EU accession. After a phase of adjustment to the fiercer competition in the single market, Austria's current account position improved.
- *EU net budget position*: Austria, as the second richest country in the EU (measured by GDP per capita in PPS), is of course a net payer into the EU budget. On average, over the period 1995–2015, it contributed 0.25 % GDP more to the EU budget than it received in transfers out of the EU budget (see Appendix 2).
- *Net migration*: This effect was rather modest vis-a-vis the EU. After German unification, an increasing number of workers from Germany entered the Austrian labor market. In our simulation, we considered (exogenously) the amount of net migration above the normal trend and interpreted this development to have been caused by EU accession (see Appendix 2).

3.1.3 EMU Membership in 1999

Participating in the EMU and thus introducing the euro further deepened economic integration. Prior to EMU membership, the hard currency countries Germany and Austria suffered from international competitiveness insofar as the soft currency countries (in the periphery of the EU) depreciated their currencies against the DM bloc in every case of current account deterioration. Of course, a devaluation race was a permanent menace for the single market. After the introduction of the euro, this was no longer possible and hence the international competitiveness was reversed within the euro area. Germany and Austria gained in the form of real depreciation, whereas the others revaluated and lost competitiveness. In addition to this advantage in the competitiveness of the formerly hard currency countries, a single currency eliminates exchange rate uncertainties, thus stimulating trade and FDI. Above all, the deeper financial integration offered new growth-enhancing stimuli via TFP-stimulating R&D growth. In scenario 3, therefore, the following three effects are considered:

- *Greater competitiveness*: The improvement of competitiveness of Austria as described above is captured by the assumption that the 1999 EMU membership led to a cessation of real appreciation (see Appendix 2).
- *Trade and FDI*: The euro's pro-trade effect—described in the theoretical part above—is captured by the 'smart dummy' variable 'Regime change T+FDI.' In the simulations, the dummy variable was increased from 1.2 to 1.3 in 1999.
- *TFP-stimulating R&D expenditures*: In addition to and on top of the growth effect of the participation in the EU's single market, the participation in the EMU is also assumed to have stimulated TFP and hence real GDP growth via an additional increase in R&D because of the stronger participation in EU research programs. In the simulations, the R&D dummy was increased from 1.1. to 1.2 in 1999.

3.1.4 EU Enlargement in 2004/2007

As a member of the EU, Austria also benefited from the major enlargement moves in 2004 and 2007, primarily because this involved mainly former Central and Eastern European countries (CEECs) in Austria's neighborhood. Two main effects were encountered: with the abolition of border controls, Austria was able to increase its trade potential in addition to the effects already happening as a result of the opening up of Eastern Europe in 1989.[6] Integration of low-income countries into the group of high-income countries in the old EU naturally induced factor movements in both directions: FDI from the West to the East, and labor migration the other way round. To mitigate the negative effects on the labor markets, many old EU member states, including Austria, applied exemption rules from freedom of labor in the form of 7-year transitional arrangements. These transition periods were phased out in the

[6]Prior to EU accession, candidate countries of the 2004 and 2007 enlargements had already abolished trade tariffs with the old EU member states in the context of the asymmetric liberalization process of the Europe agreements (EAs): The EU eliminated tariffs and NTBs on imports from the CEECS in 1997, and the CEECs did so by 2002. After EU accession, the new member states entered the customs union of the EU and participated in the EU's single market program. That meant, on the one hand, adjustments of the national external tariff to EU's CET and the abolition of border controls. Hence, the remaining trade costs were eliminated.

first round of enlargement in 2011 and in 2014 for the second round (Bulgaria, Romania).

In scenario 4, we consider only two integration effects: trade and FDI, and net migration:

- *Trade and FDI*: These effects are captured in the 'smart dummy' variable 'Regime change T+FDI.' Starting in 2004, this dummy variable was increased in the simulations from the former EMU membership value of 1.3 to 1.4; in 2007, due to the next round of EU enlargement, it was increased to 1.5.
- *Net migration*: In spite of the 7-year transitional exemptions, Austria was already attracting many specialized workers at the start of the fifth EU enlargement in 2004. Therefore, in the simulations we implemented (exogenously) positive net migration inflows above the normal trend from 2004 (see Appendix 2).

3.2 Model Results

The econometrically estimated integration model for Austria (Appendix 1) was implemented using EViews 7.0 for the period 1960–2015. The main data source is the AMECO database by the European Commission, including the latest forecast to 2015. The net migration data are provided by Statistik Austria.

3.2.1 Opening Up of Eastern Europe in 1989

The fall of the Iron Curtain was a windfall gift for the Austrian economy. This historic event provoked the already existing Habsburg or 'k.u.k Monarchy' effect. Austria quickly used these new opportunities for trade and FDI. In the model simulations (Fig. 2 and Table 2), the opening up resulted in an increase in real GDP of 0.2 percentage points per year. This created additional jobs and reduced unemployment. The current account position improved.

3.2.2 EU Membership in 1995

The main step in EU integration was when Austria became an EU member. The full exploitation of the integration effects of participation in the single market resulted in an increase in real GDP of 0.6 percentage points per year. Due to fiercer competition, inflation went down. Furthermore, 12,000 jobs could be created per year and thus unemployment decreased considerably. However, due to the confrontation with strong competitors from the old EU member states, the current account deteriorated.

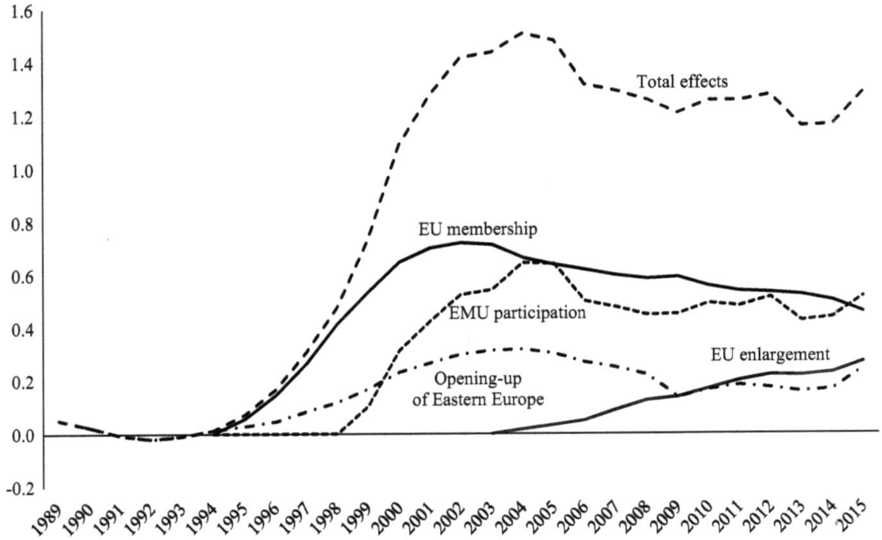

Fig. 2 Effects of Austria's participation in all steps of EU integration from 1989 (GDP, volume, percentage changes per annum)

3.2.3 EMU Membership in 1999

In addition to EU membership, the participation in the EMU and adoption of the euro added a further 0.5 percentage points per year to real GDP. Our results are similar but somewhat below those of McKinsey Germany (2012). Accordingly, Austria benefited the most from the euro (7.8 % more real GDP growth over a 10-year period, or 0.8 % per year), followed by Finland (6.7 %), Germany (6.4 %), and the Netherlands (6.2 %). The eurozone has gained 3.6 % in 10 years. The McKinsey study evaluates four categories of euro effects: (1) reduction in transaction costs (low effects on GDP); (2) intra-euro area trade effects; (3) competitiveness (this effect is high for Germany and also (as in our model) in Austria; it is negative for the soft-currency countries, such as Italy), and (4) the interest rate effect (this effect is low for Germany and Austria because the common interest rate of the euro area was based on that of Germany; it was high for the countries with high pre-EMU interest rates, such as Italy, and other countries in the euro area periphery).

3.2.4 EU Enlargement in 2004/2007

EU enlargement complemented the already ongoing advantages from the opening up of Eastern Europe for Austria. Real GDP could be increased additionally by 0.2

Table 2 Effects of Austria's participation in all steps of EU integration from 1989. Selected macroeconomic indicators

	Real GDP		CPI (inflation)	Employment total		Unemployment Rate	Unemployment Absolute	Current account balance	Budget balance
	Percent	Bn EUR 2005 prices	Percent	Percent	in 1000	Percentage points	in 1000	As a percentage of GDP	
Opening-up of Eastern Europe 1989—25 years									
1989–2015									
Cumulated	4.72	12.73	0.40	2.02	84.91	−0.52	−20.11	2.10	0.25
p.a.	0.18	0.49	0.02	0.08	3.27	−0.02	−0.77	0.08	0.01
EU Membership 1995—20 years									
1995–2015									
Cumulated	12.72	31.84	−5.07	6.06	244.7	−1.36	−49.41	−10.5	0.56
p.a.	0.58	1.59	−0.25	0.30	12.23	−0.07	−2.47	−0.53	0.03
EMU Membership 1999—15 years									
1999–2015									
Cumulated	9.30	24.00	0.80	3.86	159.18	−1.00	−38.05	−4.06	0.55
p.a.	0.53	1.50	0.05	0.24	9.95	−0.06	−2.38	−0.25	0.03
EU enlargement 2004 and 2007—10 years									
2004–2015									
Cumulated	2.44	6.71	−0.02	1.07	45.49	−0.17	−5.53	0.02	0.31
p.a.	0.20	0.61	0.00	0.10	4.14	−0.02	−0.50	0.00	0.03
Overall integration effects since 1989—25 years									
1989–2015									
Cumulated	28.55	62.65	−4.45	12.64	480.43	−2.70	−92.70	−10.23	1.44
p.a.	0.94	2.41	−0.17	0.49	18.48	−0.10	−3.57	−0.39	0.06

percentage points per year. Most studies on EU enlargement find a 1:10 rule. This means that the welfare gains of the newcomers are 10 times higher than those of the incumbent EU member states (see Breuss 2002; similarly, Levchenko and Zhang 2012[7]).

3.3 Overall Effects of Austria's EU Integration Since 1989

Due to the processes of the opening up of Eastern Europe, EU accession, EMU membership, and EU enlargement running in parallel, the integration effects of the different stages partly overlap. Hence, the various integration effects do not simply add up. All in all (see Table 2), the integration stages considered here accelerated the growth in real GDP (and only marginally less real GDP per capita) in Austria by 0.9 percentage points per year (equivalent to €2.4 billion at 2005 prices), and created around 18,000 jobs each year.[8] The unemployment rate shifted downwards by 0.1 percentage points per year and the rate of inflation by 0.2 percentage points. The ratio of imports to GDP increased altogether more than the export ratio. The entire integration process led to a weaker current account balance, mainly brought about by EU membership and EMU participation, but partly offset by the opening up of Eastern Europe. The latter factor and EU enlargement improved Austria's opportunities to participate actively in the process of globalization (or in 'mini-globalization' with regard to Eastern Europe).

The trend of the simulated effects of Austria's integration into the EU shows that for each major integration step (EU membership in 1995 and EMU participation in 1999), the growth effects are high in the beginning and subside thereafter (see Fig. 2). Only in the case of the opening up of Eastern Europe is there a rather stable positive impulse on Austria's economic growth, and the integration effects of EU enlargement in 2004 and 2007 have not yet diminished. The growth effects of Austria's EU membership and EMU participation have abated, particularly in the wake of the Great Recession of 2009 and the euro area crisis.

The effects presented in Table 2 (cumulated and annual averages) blur to some extent the 'true' profile of the integration effects by suggesting that the average growth effects cited would last permanently at that level. In reality, economic

---

[7]Levchenko and Zhang (2012) estimate welfare gains due to European trade integration since 2000 in the West (average +0.14 %; Austria, with +0.39 %, is the biggest winner) and in the East (+7.94 %). The biggest winners are Estonia (+17.25 %), Latvia (+11.93 %), and Bulgaria (+10.57); the welfare gains of the other CEES are below 10 %.

[8]The detailed results for the four scenarios, and also the overall results, are quite similar to those of the earlier study, which covered the period 1989–2011 (see Breuss 2012, 2013c).

integration, i.e., the accession of a country to an integrated community (EU), gives rise to initial positive growth incentives (mainly due to a necessary adjustment and productivity shock), which gradually fade. We therefore observe, as a rule, 'falling marginal returns' for integration. Even after the growth effects have faded away, the level of income (real GDP) has been raised cumulatively by 29 % (or by €63 billion at 2005 prices) as a result of participation in all integration steps over the 25 years since the opening up of Eastern Europe. At the same time, real GDP per capita (welfare) increased cumulatively by 28 % or by €7000. However, the welfare gain brought about by participation in European integration is defined not only by the level and growth of GDP per capita: it also includes the increase in the variety of goods and services supplied and options for individual action (free movement and the Schengen Agreement facilitate labor mobility and travel, the latter also benefiting from the common currency), as well as the modernization of the political system by introducing the European dimension. Moreover, full participation in the EU single market implies permanent downward pressure on prices and raises private household purchasing power. This effect is prolonged, and is reinforced by each round of EU enlargement and the accompanying extension of the single market. In contrast to the suggestion by some authors of the new growth theory of foreign trade, integration is found to have no permanent effects on growth rates, but provides one-off incentives for economic growth, which raise the level of GDP initially but ebb thereafter.

According to the calculations in this study, Austria has benefited economically from all stages of integration (opening up of Eastern Europe, EU membership, EMU participation, and EU enlargement). The integration effects derived from model simulations for Austria's participation in all EU integration moves are in the order of ½ to 1 percentage point of additional GDP growth per year. However, not all parts of the Austrian economy have profited equally from EU integration. The primary winners are those companies heavily engaged in the new EU member states. An indication is the decline in the wage share since the 1980s. The 'mini-globalization' has clearly exerted pressure on wages (see Breuss 2010c).

The plausibility of these model results is confirmed when Austria's economic performance is compared to that of other countries inside or outside the EU: Austria's growth advantage vis-à-vis Germany and Switzerland roughly corresponds to the above-cited integration effects. This 'growth dividend' is difficult to explain—if it can be explained at all—when abstracted from the integration effects of Austria's participation in all EU policy moves.

4 Conclusions

The euro area crisis has confronted the EU with new challenges. The previous governance architecture of the EMU did not withstand the test of the crisis. To prevent the euro area from breaking up, the governance of the EMU has been readjusted to make it more resilient to future shocks. High on the agenda are the convergence of competitiveness among euro countries (monitored and steered by the new procedure for 'excessive macroeconomic imbalances' within the framework of the Six-Pack and Two-Pack reforms—ideally heading towards a homogeneous European business cycle), and in particular the longer-term reduction of the (in some periphery countries) unsustainably high government debt, coupled with the containment of the debt dynamics through the instruments of the Six Pack (reform of the Stability and Growth Pact) and accompanying measures provided for by the Fiscal Compact (e.g., debt brakes at the national level). Beyond the tools for the closer coordination and centralization of fiscal policy, the EU—and notably the euro area—have the European Stability Mechanism (ESM) at their disposal and are ready to embark on 'banking union' (see Breuss et al. 2015), with common bank supervision, resolution, and deposit guarantee at the EU level. Whether the EU will move even further (as suggested in the plans to reform the EMU by Barroso and Van Rompuy) in the direction of centralization ('political union' or the 'United States of Europe') is still open (for more, see Breuss 2013a, b). For some member countries, such development may go too far and would provoke their early withdrawal (e.g., the UK), or deepen the rifts within the EU and the euro area that have emerged since the crisis.

In any case, historical studies on the reduction of public debt do not bode well for Europe from a medium- and longer-term perspective. All measures to slash government debt by means of fiscal austerity (expenditure cuts and tax increases, as foreseen by the Six-Pack reforms and the Fiscal Compact) may dampen medium- and long-term economic growth (see the extremely negative results in Greece).

Due to these negative perspectives, the 'growth dividend' that Austria has enjoyed in the past, benefiting from its strong involvement in Eastern European 'emerging markets,' may gradually wane. As already signaled by current medium-term projections, the new member countries in Eastern Europe may also move onto a slower growth path, as they will be affected indirectly by the euro area crisis and the negative side effects of its resolution (notably the collective de-leveraging), and the ad hoc break-out of political crises, such as that in the Ukraine, and the ensuing tensions between the EU and Russia.

Appendix 1: The Estimated Integration Model for Austria

Real GDP (Cobb–Douglas production function; bn. EUR, 2005 prices) GDPR = (TFP) * ((K^0.26) * (EE^0.74))
Total factor productivity (TFP) DLOG(TFP) = −0.0117597194657 + 0.975350400527 * DLOG(AP) + 0.00368866066045 * RAD + 0.000364739422324 * D(XQUOTA)
Research & development: R&D in % of GDP RAD = −0.771758304314 + 0.0900123360683 * LOG(GDPR) + 0.918022689413 * RAD(−1) + 0.450963636885 * D_1995_2015
Private consumption deflator DLOG(PCN) = 0.974494644295 * DLOG(CPI) − 0.0100090054202 * D_2002
Private consumption index: National definition DLOG(CPI) = 0.00685148354097 + 0.210308218697 * MARKUP * DLOG(ULC) + 0.232379177613 * MARKUP * DLOG(PM) + 0.407094518941 * DLOG(CPI(−1)) + 0.014977340126 * D_1984
Harmonized index of consumer prices: HICP DLOG(HICP) = 0.974397164556 * DLOG(CPI)
GDP deflator DLOG(PGDP) = 0.883285761406 * DLOG(CPI) + 0.432199804412 * DLOG(PX) − 0.275658593485 * DLOG(PM)
Wage per employee (Phillips curve) DLOG(WE) = −0.00688732197519 + 0.658922532489 * DLOG(CPI) + 0.439378457835 * DLOG(AP(−1)) + 0.0643927893279 * 1 / U − 0.059812308921 * D_1980
Wages WN = (WE * E) / 1000
Taylor rule for the euro area RSH_EA18 = 2 + DLOG(HICP_EA18) * 100 + 0.5 * (DLOG(HICP_EA18) * 100 − 2.0) + 0.5 * (DLOG(GDPR_EA18) * 100 − 1.5)
Interest rate, short-term RSH = −5.5262147236 + 0.667905535844 * RSH_EA18 + 0.0136803208004 * LOG(CPI) * 100 − 2.35378845633 * D_1983
Interest rate, long-term RLH = 0.400105575997 + 0.23428902887 * RSH + 0.152448613707 * DLOG(CPI) * 100 + 0.674036427571 * RLH(−1)
Capital demand DLOG(K) = 0.000347597990373 + 0.000690816072569 * D(BUD) − 0.000555726084856 * PRDEF − 0.10749650936 * DLOG(GDPR) + 0.000211122052822 * D(DLOG(WE) * 100 − (RLH − DLOG(PGDP) * 100)) + 0.879984556303 * DLOG(K(−1))

(continued)

Capital coefficient: K/Y KY = (K / GDPR)
Labor demand (total employment) DLOG(EE) = 0.174447800692 * DLOG(GDPR) − 0.0646686954094 * DLOG(WE) + 0.00183780684966 * D(BUD) + 0.688076954685 * DLOG(EE(−1))
Labor demand (employees) DLOG(E) = −0.0020926578709 + 0.787853784774 * DLOG(EE) + 0.174748348465 * DLOG(GDPR) + 0.262099988497 * DLOG(E(−1))
Labor supply: Labor force LS = EE + US
Labor productivity (total economy) AP = (GDPR / EE)
Unit labor costs ULC = (WN / GDPR)
Unemployment rate (Okun's law) D(U) = 0.0856028080042 − 7.48943374025 * DLOG(GDPR) + 0.00304288354196 * D(POP − MIGR_OST89 − MIGR_EU95 − MIGR_EUEW04) + 0.804600244209 * D_1982 − 0.0362182637141 * BUD
Unemployment, total in 1000 persons US = ((U * LS) / 100)
Exports of goods and services, total, real DLOG(XGSR) = −0.0436572302437 + 2.22907387142 * DLOG(GDPR_EU28) − 0.555430829575 * DLOG(REER_IC37) + 0.0393558155438 * D_1989_2015
Exports of goods and services, total, nominal bn. EUR XGSN = XGSR * (PX / 100)
Export quota: Exports goods and services in % of GDP XQUOTA = (XGSN / GDPN) * 100
Imports of goods and services, total, real LOG(MGSR) = −5.3567516112 + 1.77756769413 * LOG(GDPR) + 0.228751889216 * D_1989_2015
Imports of goods and services, total, nominal bn. EUR MGSN = MGSR * (PM / 100)
Import quota: Imports goods and services in % of GDP MQUOTA = (MGSN / GDPN) * 100
Current account in nominal bn. EUR (AMECO) CA = XGSN − MGSN
Current account in % of GDP (AMECO) CAGDPN = ((XGSN − MGSN) / GDPN) * 100

(continued)

Current account in nominal bn. EUR (OeNB) CA_OeNB = CA − CA_Diff_to_OeNB
Current account in % of GDP (OeNB) CA_OeNBGDPN = ((CA_OeNB) / GDPN) * 100
FDI outflows in % of GDP FDIEX = 0.375640070717 + 1.02837425753 * D(FDISOUT)
FDI outward stocks in % of GDP FDISOUT = −23.7147058544 + 0.883784157118 * FDISOUT(−1) + 23.3682906272 * D_1989_2015
FDI inflows in % of GDP FDIIN = 0.671986218682 + 0.84990945751 * D(FDISIN)
FDI inwards stocks in % of GDP FDISIN = −28.0471754242 + 0.810412880324 * FDISIN(−1) + 28.0293244537 * D_1989_2015
Net household disposable income, nominal (bn. EUR; OECD Economic Outlook; Macrobond) YDN = 2.18851454149 + 0.11686303161 * GDPN + 0.817157924902 * YDN(−1)
Net household disposable income, real (bn. EUR) YDR = (YDN / (PCN / 100))
GDP, nominal (bn. EUR) GDPN = (GDPR * (PGDP / 100))
Real GDP per capita (in 1000 EUR) – WELFARE measure 1 GDPRPC = ((GDPR * 1000) / (POP − MIGR_OST89 − MIGR_EU95 − MIGR_EUEW04))
GDP per capita in PPS (EU-28 = 100) – WELFARE measure 2 LOG(GDPPC_PPSEU28) = 0.43328354923 + 0.00346210004573 * DLOG(GDPRPC) + 0.911257550549 * LOG(GDPPC_PPSEU28(−1)) − 0.0461887756332 * D_2001
Budget position: Budget balance in % of GDP BUD = −1.28851868518 + 0.354920098741 * DLOG(GDPR) * 100 − 0.594239170511 * ELEC + 0.700806989349 * BUD(−1) − 2.70112458588 * D_2004
Budget position: Budget balance in % of GDP incl. Net contribution to EU budget BUDNET = BUD + NETEU
Austria–EU Budget position absolute values in bn. EUR NETEUABS = (NETEU * GDPN) / 100
Public debt dynamics: Gross public debt in % of GDP (DEBT = DEBT(−1) − PD − (r − g)*DEBT(−1) + SF (Stock flow)) DEBT = DEBT(−1) − PRDEF + SNOW + SF
Primary budget balance in % GDP PRDEF = BUD − INTEREST

(continued)

Interest payments in % of GDP INTEREST = 0.187508058025 + 7.27693766331 * (RLH / 100) * ((DEBT(−1)) / GDPN(−1)) + 0.893137557651 * INTEREST(−1)
Snow-ball effect SNOW = 0.276597903339 + 0.00796005959488 * (RLH − DLOG(GDPN) * 100) * DEBT(−1)
Wage share: Wages in % of GDP ('Globalization' reduces LQ) LQ = 15.1699479237 − 0.0316886728056 * (XQUOTA + MQUOTA) − 0.00942994300939 * D(FDISOUT + FDISIN) + 0.791815264509 * LQ(−1) + 3.87639625065 * D_1975

DLOG(Variable) = percentage change operator. Estimation with EViews 7.0 for the period 1960–2015. Data source AMECO database of the European Commission; PX (PM) = deflators of exports (imports) of goods and services; D_1989_2015 = 'smart' dummy 'Regime change T+FDI'; D_1995_2015 = 'smart' dummy for 'Regime change R&D'; FDI = foreign direct investment; OeNB = Austrian National Bank

Appendix 2: Quantitative Model Inputs of Four Integration Scenarios (Additional Effects Compared to the Baseline Scenario Without EU Integration)

	Scenario 1		Scenario 2					Scenario 3			Scenario 4	
	Open-1989		EU-1995					EMU-1999			EU-Enlarg-2004/07	
	T+FDI	MIGR	Mark-up	R&D	T+FDI	EU-Budg	MIGR	REER	T+FDI	R&D	T+FDI	MIGR
1989	1.1	40	1.3	1.0	1.1	0	0	103.91	1.1	1.0	1.1	0
1990	1.1	55	1.3	1.0	1.1	0	0	103.07	1.1	1.0	1.1	0
1991	1.1	73	1.3	1.0	1.1	0	0	102.55	1.1	1.0	1.1	0
1992	1.1	67	1.3	1.0	1.1	0	0	103.77	1.1	1.0	1.1	0
1993	1.1	30	1.3	1.0	1.1	0	0	110.06	1.1	1.0	1.1	0
1994	1.1	0	1.3	1.0	1.1	0	0	112.70	1.1	1.0	1.1	0
1995	1.1	0	1.2	1.1	1.2	−0.44	−2	114.80	1.2	1.1	1.2	0
1996	1.1	0	1.1	1.1	1.2	−0.15	0	113.43	1.2	1.1	1.2	0
1997	1.1	0	1.0	1.1	1.2	−0.43	−2	108.09	1.2	1.1	1.2	0
1998	1.1	0	1.0	1.1	1.2	−0.34	4	106.82	1.2	1.1	1.2	0
1999	1.1	0	1.0	1.1	1.2	−0.32	15	107.00	1.3	1.2	1.3	0
2000	1.1	0	1.0	1.1	1.2	−0.21	13	107.00	1.3	1.2	1.3	0
2001	1.1	0	1.0	1.1	1.2	−0.26	33	107.00	1.3	1.2	1.3	0
2002	1.1	0	1.0	1.1	1.2	−0.10	29	107.00	1.3	1.2	1.3	0
2003	1.1	0	1.0	1.1	1.2	−0.15	36	107.00	1.3	1.2	1.3	0
2004	1.1	0	1.0	1.1	1.2	−0.16	14	107.00	1.3	1.2	1.4	33
2005	1.1	0	1.0	1.1	1.2	−0.11	14	107.00	1.3	1.2	1.4	26
2006	1.1	0	1.0	1.1	1.2	−0.12	14	107.00	1.3	1.2	1.4	6
2007	1.1	0	1.0	1.1	1.2	−0.21	14	107.00	1.3	1.2	1.5	7
2008	1.1	0	1.0	1.1	1.2	−0.13	14	107.00	1.3	1.2	1.5	7
2009	1.1	0	1.0	1.1	1.2	−0.15	14	107.00	1.3	1.2	1.5	0

(continued)

	Scenario 1		Scenario 2					Scenario 3			Scenario 4	
	Open-1989		EU-1995					EMU-1999			EU-Enlarg-2004/07	
	T+FDI	MIGR	Mark-up	R&D	T+FDI	EU-Budg	MIGR	REER	T+FDI	R&D	T+FDI	MIGR
2010	1.1	0	1.0	1.1	1.2	−0.24	14	107.00	1.3	1.2	1.5	3
2011	1.1	0	1.0	1.1	1.2	−0.27	14	107.00	1.3	1.2	1.5	13
2012	1.1	0	1.0	1.1	1.2	−0.35	14	107.00	1.3	1.2	1.5	26
2013	1.1	0	1.0	1.1	1.2	−0.35	14	107.00	1.3	1.2	1.5	20
2014	1.1	0	1.0	1.1	1.2	−0.35	14	107.00	1.3	1.2	1.5	30
2015	1.1	0	1.0	1.1	1.2	−0.35	14	107.00	1.3	1.2	1.5	30

Scenarios: 1 = opening up of Eastern Europe 1989; 2 = EU membership 1995; 3 = EMU membership 1999; 4 = EU enlargement 2004/2007; T+FDI = dummy for 'regime change' in trade and FDI (original values: 1 until 1988; starting with 1989, 0.1 points higher with each integration step: 1989 = 1.1; 1995 = 1.2; 1999 = 1.3; 2004 = 1.4; 2007 = 1.5; MIGR = net migration (in 1000 persons) due to respective integration step; Mark-up = dummy for mark-up pricing, decreasing due to fiercer competition when participating in the EU's single market; R&D = dummy for 'regime change' in research & development (R&D) policy due to EU/EMU membership; EU-Budg = net payer position (in % of GDP); REER = real effective exchange rate (entering EMU has stopped the previous trend of appreciation; i.e., the increase in REER)

References

Badinger H, Breuss F (2005) Has Austria's accession to the EU triggered an increase in competition? A sectoral markup study. Empirica 32(2):145–180
Badinger H, Breuss F (2011) The quantitative effects of European post-war economic integration. In: Jovanovic MN (ed) International handbook on the economics of integration: factor mobility, agriculture, environment and quantitative studies, vol III. Edward Elgar, Cheltenham, pp 285–315
Baldwin R, Venables AJ (1995) Regional economic integration. In: Grossman GM, Rogoff K (eds) Handbook of international economics, vol III. Elsevier Science B.V., Amsterdam, pp 1597–1644
Breuss F (2002) Benefits and dangers of EU enlargement. Empirica 29(3):245–274
Breuss F (2003a) Reale Außenwirtschaft und Europäische Integration. Peter Lang – Europäischer Verlag der Wissenschaften, Frankfurt am Main
Breuss F (2003b) Austria, Finland and Sweden in the European Union: economic effects. Austrian Econ Q 4:131–158
Breuss F (2006) Monetäre außenwirtschaft und Europäische integration. Peter Lang – Europäischer Verlag der Wissenschaften, Frankfurt am Main
Breuss F (2010a) 15 years of Austrian EU membership. Austrian Econ Q 15:165–183
Breuss F (2010b) An evaluation of the EU's fifth enlargement: with special focus on Bulgaria and Romania. In: Szekeley I, Keereman F (eds) Five years of an enlarged EU – a positive sum game. Springer, Berlin, pp 221–248
Breuss F (2010c) Globalisation, EU enlargement and income distribution. Int J Public Policy 6(1–2):16–34
Breuss F (2012) EU-Mitgliedschaft Österreichs. Eine Evaluierung in Zeiten der Krise. WIFO, Vienna, October 2012. Available from: http://www.wifo.ac.at/wwa/pubid/45578
Breuss F (2013a) Towards a new EMU. WIFO Working Papers, Nr. 447, March 2013
Breuss F (2013b) Towards United States of Europe. In: Visions for economic policy coordination in Europe. Federal Ministry of Economy, Family and Youth, Vienna, pp 27–47
Breuss F (2013c) Effects of Austria's EU membership. WIFO Austrian Econ Q 18:103–114, Available from: http://www.wifo.ac.at/jart/prj3/wifo/main.jart
Breuss F (2014) A prototype model of European integration: the case of Austria. WIFO Working Papers, No. 465, March 2014
Breuss F, Kratena K, Schebeck F (1994) Effekte eines EU-Beitritts für die Gesamtwirtschaft und für die einzelnen Sektoren, Sonderheft der WIFO-Monatsberichte Österreich in der Europäischen Union: Anforderungen und Chance für die Wirtschaft, Wien
Breuss F, Roeger W, Veld J (2015) The stabilising properties of a European Banking Union in case of financial shocks in the euro area. European Economy, Economic Papers, No. 550, European Commission, Brussels, June 2015
Felbermayr G, Larch M, Krüger F, Flach L, Yalcin E, Benz S (2013) Dimensionen und Auswirkungen eines Freihandelsabkommens zwischen der EU und den USA, ifo Forschungsbericht, Nr. 62, München 2013
Hamilton JD (2008) Regime-switching models. In: Durlauf SN, Blume LE (eds) The New Palgrave dictionary of economics, 2nd edn. Palgrave Macmillan. Available from: http://www.dictionaryofeconomics.com/article?id=pde2008_R000269. doi:10.1057/9780230226203.1411
Jovanovic MN (ed) (2011) International handbook on the economics of integration (3 vols). Edward Elgar, Cheltenham
Keuschnigg C, Kohler W (1996) Austria in the European Union: dynamic gains from integration and distributional implications. Econ Policy 22:155–211

Levchenko AA, Zhang J (2012) Comparative advantage and the welfare impact of European integration. NBER Working Paper, No. 18061, May 2012
McKinsey Germany (2012) The future of the euro: an economic perspective on the eurozone crisis. McKinsey & Company, Frankfurt
Viner J (1950) The customs union issue. Carnegie Endowment, New York

Trade Agreements and Regional Disparities

Pasquale Commendatore, Ingrid Kubin, and Iryna Sushko

1 Introduction

The strengthening of the EU internal integration is a well-documented fact and three periods are distinguished in Dorrucci et al. (2002). The first started in March 1957 with the Treaty of Rome, where the period included the implementation of the Common Agricultural Policy in 1962 and it ended in July 1968 when the Customs Union was completed. This period was characterized by rapid integration, which slowed down considerably in a second period between the early 1970s and the mid-1980s, with the notable exception of the start of the European Monetary System in March 1979. In the third period, the creation of the Common Market and the Monetary Union again led to the considerable acceleration of regional integration. As a result, in the early 2000s, the EU/euro area could already be classified as somewhere between an economic union and total economic integration, which are the highest levels of economic integration differentiated in Dorrucci et al. (2002). This integration process reduced barriers to commodity trade as well as barriers to factor mobility.

However, the EU is still characterized by deep differences within and across nations. According to the Eurostat regional yearbook 2010, the GDP per inhabitant

P. Commendatore
University of Naples Federico II, Naples, Italy
e-mail: commenda@unina.it

I. Kubin (✉)
WU Vienna University of Economics and Business, Vienna, Austria
e-mail: ingrid.kubin@wu.ac.at

I. Sushko
National Academy of Science of Ukraine, Institute of Mathematics, Kyiv, Ukraine

Kyiv School of Economics, Kyiv, Ukraine

B. Bednar-Friedl, J. Kleinert (eds.), *Dynamic Approaches to Global Economic Challenges*, DOI 10.1007/978-3-319-23324-6_3

of 67EU-27 NUTS 2 regions, out of 271, is below 75 % of the average, whereas that in 41 regions is greater than 125 % of the average. These regional disparities are significant among, but also within countries (especially in Turkey and Slovakia, but also in Italy, Germany, and the UK).

In this study, we consider the economic effects of intensified trade integration within the Union in which productive factors are also mobile. We argue that a New Economic Geography (NEG) perspective is preferable to an Heckscher-Ohlin approach. We show that trade integration leads to specialization, trade creation, and trade diversion (as predicted by a standard Heckscher-Ohlin framework), but it may also lead to agglomeration within the Union, which is typical of a NEG framework. We show that these agglomeration processes reinforce the specialization and trade effects of trade integration. Finally, we provide insights into the dynamic processes and we find that the coexistence of attractors in our modeling approach is more pervasive than in a standard NEG framework.

2 Modeling Strategy and Related Studies

2.1 Heckscher-Ohlin or NEG?

The traditional analytical framework for assessing the effects of trade agreements is a three-country Heckscher-Ohlin model, where two countries form a Union to reduce their mutual tariffs, while still maintaining the tariff with respect to the third country. Trade increases between the member countries because these countries increase their specialization and they import commodities that they previously produced themselves (trade is created), and because the member countries import commodities from other member countries, which they previously imported from a country outside the Union (trade is diverted).

A Heckscher-Ohlin framework assumes that countries differ with respect to relative factor endowments, markets are fully competitive, and productive factors are immobile between countries. However, many trade agreements (the treaty of Rome is a prominent example) are actually formed between countries that are similar with respect to factor endowment, and productive factors are free to move between the member countries. In addition, production is often characterized by decreasing average costs, products are differentiated and thus markets are monopolistically competitive.

Therefore, in the following, we analyze the effects of trade agreements in a NEG framework that permits all of these elements. Standard NEG models, which were originally proposed by Krugman (1991), are concerned mainly with the effects of trade integration on the distribution of economic activity between two countries that belong to the same integrated economic area. The standard long-run outcomes involve spreading the economic activity across space, or industrial concentration in one of the two countries, depending on the relative strength of agglomeration and dispersion forces. The first type of force depends on the proximity of firms and households to the larger market, whereas the second depends on the competition in goods and factors markets, which occur as firms become more concentrated in one location. In a two-country framework, increasing trade openness typically favors agglomeration to dispersion forces and long-term spatial concentration may emerge. However, our question requires that we extend the standard model to a three countries model where two countries are within the Union and one country is outside.

2.2 *Isoelastic or Linear Demand Functions?*

In previous NEG studies, only a few models have considered the analysis of integration areas. The simplest modeling structure involves a trade area (the Union in our problem) formed of two symmetric members, which are equally (and increasingly) accessible from an outside country. It should be noted that most of these analyses concentrated on external trade liberalization, whereas we focus on internal integration. Interestingly, the prediction concerning the impact of external liberalization on internal inequalities is not independent of the modeling specification selected (see Brülhart 2011; Commendatore et al. 2015). In general, previous studies can be grouped into two classes, as follows. In the first class of models, external trade liberalization leads to the prevalence of agglomeration over dispersion forces (see Paluzie 2001; Brülhart et al. 2004; Commendatore et al. 2014). In these models, which are characterized by isoelastic demand functions, multiplicative (iceberg) trade costs, and fixed mark-ups, greater external trade openness reduces the strength of internal competition more than the importance of proximity to the larger internal market. The overall effect is similar to a reduction of internal trade costs, and a shorter distance from the external country fosters increasing inequalities within the Union, thereby leading to agglomeration (i.e., to the so-called core periphery equilibrium). By contrast, in the second class of models, trade liberalization may lead to the prevalence of dispersion over agglomeration forces. In these class of models, a further dispersion force operates: In Krugman and Livas Elizondo, 1996, it is represented by increases in land rentals and commuting costs following worker migration. Instead, in the model of Behrens (2011), which is characterized by linear demand functions, additive trade costs, and variable mark-ups, external trade liberalization strengthens the competition effect by reducing local mark-ups. Due to this additional competition effect, external trade liberalization may actually lead to convergence across countries within the Union.

In our subsequent analysis, we employ the latter specification with linear demand functions for two reasons. First, we consider the additional competition effect via a variable mark-up as relevant. Second, we expect that this setup will permit more analytical results. The analysis of NEG models typically focuses on the regional pattern of industrial activity and little attention is paid to sectoral specialization, trade creation, and trade diversion, although these questions are at the core of the standard (Heckscher-Ohlin) approach to integration areas. Commendatore et al.

(2014) approached those questions in a NEG framework with an isoelastic demand function and showed that trade liberalization also affects the sectoral employment structure of the two members of the Union, although the analytic complexity of the model limited the scope for obtaining explicit analytic results. Therefore, we use a model with linear demand functions in the following.

2.3 *Which of the Productive Factors is Mobile?*

In contrast to Heckscher-Ohlin models where productive factors are assumed to be immobile, a prototype NEG model accounts explicitly for factor mobility. How this mobility process is modeled is a core feature that actually defines the different NEG model classes. Multi-region NEG models typically assume that factor mobility is free within the Union whereas it is restricted outside the Union. Many models are based on labor mobility as a dynamic process that brings about agglomeration (such as: Krugman and Livas Elizondo 1996, with respect to Mexico; Paluzie 2001, with respect to the Spanish industrialization process; Wang and Zheng 2013a and Wang and Zheng 2013b with respect to China) where labor mobility is plausible in all cases. However, high labor mobility is less plausible in the context of the European Union. In a study of the EU regions at the NUTS 2 level, Gakova and Dijkstra (2008) showed that the average share of working age residents moving in another EU region represents less than 1 % of the EU's working age population (compared with 2 % in the US). Migration in Europe is rather weak, so as far as the EU is concerned, the mobility of unskilled workers does not actually appear to play a role in the adjustment process to wage differentials among countries (Siebert 1997; Obstfeld and Peri 1998; Puga 2002). By contrast, firm mobility has assumed an increasing role since the EU enlarged to include Eastern European countries from the mid-1990s.

Therefore, we employ a model framework where firms and entrepreneurial, highly qualified human capital is mobile within the Union, whereas (unqualified) labor is immobile (similar to Brülhart et al. 2004). Note that the differentiation between two types of labor resembles the Heckscher-Ohlin perspective.

Thus, we utilize a three-country footloose entrepreneur model, which we analyze in the following sections. As is typical of a NEG analysis, we specify the fundamentals in the first step: factor endowments, technology and utility. In addition, a NEG model also includes a rudimentary geographical structure in terms of specifying costs for shipping commodities between countries. In the second step, we analyze the short-run equilibrium, which is characterized by profit and utility maximization, and by market clearing for a *given* regional allocation of the mobile factor. Typically, factor rewards will differ between countries in the short run. The third step, which is characteristic of a NEG perspective, investigates the factor mobility process triggered by these differences in factor rewards and its long-run implications. In particular, will all productive factors move to one country, or will they be equally distributed over space? The third step involves a difference equation, its fixed point, and stability properties.

3 Linear Footloose Entrepreneur Model

3.1 *Fundamentals: Factor Endowments, Technology, Utility, and Geography*

This model was proposed by Ottaviano et al. (2002) in the context of a two-country economy. In the following, we formulate it explicitly for three countries,[1] where countries 1 and 2 are assumed to be symmetric and they form a Union, which is integrated with respect to factor mobility and commodity trade, whereas country 3 is an outside country. Each of the countries is endowed with qualified (entrepreneurial) labor and with unqualified labor. The total number of entrepreneurs is E, $\tilde{n}$ denotes the share of firms that are located within the Union and that are mobile between country 1 and 2; x_t and $1 - x_t$ are the corresponding shares of mobile entrepreneurs located in countries 1 and 2; and finally, $(1 - \tilde{n})$ is the share of entrepreneurs that are immobile and located in country 3. The total number of (unqualified) workers is L and θ denotes the share of workers that reside in each of the two countries within the Union; and $(1 - \theta)$ is the share of workers located in the outside country, 3. Unqualified labor is assumed to be immobile.

The economies produce two commodities. First, a homogeneous "agricultural" commodity that uses one worker per unit of output and which is traded without costs between all three countries on a competitive market. In this setup, the prices are equal to wages and they are the same in all three countries. As found typically in NEG models, we use the agricultural price/wage as numéraire (and thus it is equal to 1).

Second, the economies produce a differentiated "manufactured" commodity, where each of the variants requires one entrepreneur as a fixed input (and thus the number of variants is equal to the number of entrepreneurs) and MC workers are used per unit of output. Again, the goods are traded between all countries, but trade involves trade costs in the manufacturing sector, so shipping one unit requires T units of the agricultural good. Trade costs are assumed to be symmetric and lower for trade within the Union than for trade with the outside country. T_U denotes the trade costs within the Union, i.e., between country 1 and 2, and $T_F > T_U$ are the

[1] The structure of our model is similar to the three-country linear core periphery model proposed by Behrens, 2011.

trade costs involved in shipping the manufactured good to and from the third outside country. When appropriate, we use the following matrix notation for trade costs:

$$\mathbf{T} = \begin{bmatrix} 0 & T_U & T_F \\ T_U & 0 & T_F \\ T_F & T_F & 0 \end{bmatrix}$$

and $T_{r,s}$, with $r = 1, 2, 3$ and $s = 1, 2, 3$, which denote the costs for shipping one unit produced in country r to country s. Trade costs are central parameters in a NEG formulation, where they are broadly defined and they include direct transportation costs as well as all of the costs involved in selling a commodity in a country different from that where it was produced, in particular, they also include tariffs. Therefore, trade integration is represented by a reduction in trade costs.

This structure of the production function involving fixed inputs and constant marginal production and trade inputs (combined with the preference for product variety, as explained in the following) implies that manufactured varieties are sold in a monopolistically competitive market. Price setting leads to operating profits that remunerate entrepreneurs (and thus they also represent the fixed costs). In this specification, producers residing in one country do not differ, i.e., they are symmetric.

Consumer preferences are described by a quasi-linear utility function with two parts: a quadratic function that defines the preferences in terms the consumption c_i of the manufactured variant i ($i = 1, \ldots, E$); and a linear component that defines the consumption of the undifferentiated homogeneous agricultural good, which is denoted by C_A :

$$U = \alpha \sum_{i=1}^{E} c_i - \left(\frac{\beta - \delta}{2}\right) \sum_{i=1}^{E} c_i^2 - \frac{\delta}{2} \left(\sum_{i=1}^{E} c_i\right)^2 + C_A, \tag{1}$$

where α represents the intensity of preferences for the manufactured varieties, with $\alpha > 0$; δ represents the degree of substitutability across these varieties, with $\delta > 0$; and the difference $\beta - \delta > 0$ measures the taste for variety. Obviously, $\delta = 0$ is a limiting case with non-substitutability and a strong preference for product variety. We also analyze this case quite often because it delivers neat analytic expressions with a straightforward economic interpretation. The general case with $0 < \delta < \beta$ is also analyzed; typically, the conclusions drawn from the special case carry over (but they are less evident).

The budget constraint of an individual consumer is

$$\sum_{i=1}^{E} c_i p_i + C_A = y + \overline{C_A},$$

where p_i is the price at the destination of one industrial variety (and the price of the agricultural commodity, which is used as the numéraiere, is equal to 1), y denotes the individual factor income, and $\overline{C_A}$ is an individual endowment with the agricultural good, which is assumed to be sufficiently large to allow a positive consumption of the numéraiere in equilibrium.

3.2 Short-Run Equilibrium

The short-run equilibrium involves utility maximization, profit maximization, and market clearing for a given regional allocation of the mobile factor. In country 1, the number of entrepreneurs is $x_t \tilde{n} E$; in country 2, it is $(1 - x_t)\, \tilde{n} E$; and in country 3, it is $(1 - \tilde{n})\, E$.

Utility maximization for a customer living in s implies a linear demand function for variety i

$$p_{is,t} = \alpha - (\beta - \delta)\, c_{is,t} - \delta \sum_{i=1}^{E} c_{is,t},$$

where t is a time index; or, when solved for the quantity demanded:

$$c_{is,t} = \frac{\alpha}{(E-1)\delta + \beta} - \frac{1}{\beta - \delta} p_{is,t} + \frac{\delta}{(\beta - \delta)[(E-1)\delta + \beta]} \sum_{i=1}^{E} p_{is,t} =$$

$$= a - (b + Ec) p_{is,t} + cP_{s,t}, \tag{2}$$

where $P_{s,t} = \sum_{i=1}^{E} p_{is,t}$ is the aggregate price index in country s and a, b, and c are implicitly defined parameters, which allow the equations to be written in a more compact form, but which are less prone to an economic interpretation. When appropriate, we return to the original parameters.

Observing that producers residing in country r and selling to country s set the same price, $p_{rs,t}$ (symmetry assumption), allows us to simplify these expressions, and thus the demand of a single customer faced by a representative firm in each country is (see 2):

$$c_{rs,t} = a - (b + cN) p_{rs,t} + cP_{s,t}, \tag{3}$$

where $c_{rs,t}$ is the demand of a customer located in country s (with $s = 1, 2, 3$) for a variant produced in country r (with $r = 1, 2, 3$), and

$$P_{s,t} = x_t \tilde{n} E p_{1s,t} + (1 - x_t)\, \tilde{n} E p_{2s,t} + (1 - \tilde{n})\, E p_{3s,t.},$$

where the number of producers residing in each country is considered.

Note that these demand functions involve no income effects; therefore, aggregate demand is controlled by the number of customers, but it is independent of the income level. The market size effect still exists but it is related only to the number of customers that live in a country (and not to their incomes). Because it is linear, the demand function appears to be analytically simpler than the isoelastic counterpart used in many NEG models, but it is actually still quite complex, and thus we also analyze the special case of $\delta = 0$. The demand for the agricultural commodity is the residual demand.

Finally, the indirect utility is given by:

$$V = S + y + \overline{C}_A,$$

where S corresponds to the consumer's surplus :

$$\begin{aligned} S &= U(c_i(p_i),\ i \in [0, E]) - \sum_{i=1}^{E} p_i c_i(p_i) - C_A \\ &= \frac{a^2 E}{2b} + \frac{b + cE}{2} \sum_{i=1}^{E} p_i^2 - a \sum_{i=1}^{E} p_i - \frac{c}{2} \left(\sum_{i=1}^{E} p_i \right)^2 \\ &= \frac{a^2 E}{2b} + \frac{b + cE}{2} \sum_{i=1}^{E} p_i^2 - aP - \frac{c}{2} P^2. \end{aligned} \tag{4}$$

We now proceed to profit maximization. Each of the entrepreneurs sells to all three countries and sets specific profit-maximizing prices for the three markets subject to his cost structure and demand functions, as given in (3).

In addition, short-run equilibrium requires that demand equals supply for each variety:

$$c_{rs,t} = q_{rs,t}, \tag{5}$$

where $q_{rs,t}$ is the output produced in country r that is sold to a single customer in country s. Profit maximization leads to the following prices:

$$p_{rs,t} = \frac{a + cP_{s,t} + MC(b + cE)}{2(b + cE)} + \frac{T_{rs}}{2}. \tag{6}$$

From the assumption of the symmetric behavior of firms, it follows that prices differ across countries only because of transport costs. The price indices are:

$$P_{1,t} = \frac{a + (b + cE)\left[MC + (1 - x_t)\tilde{n}T_U + (1 - \tilde{n})T_F\right]}{2b + cE} E$$

$$P_{2,t} = \frac{a + (b + cE)\,[MC + x_t\tilde{n}T_U + (1 - \tilde{n})T_F]}{2b + cE}E$$

$$P_{3,t} = \frac{a + (b + cE)(MC + \tilde{n}T_F)}{2b + cE}E. \tag{7}$$

It is interesting to note that due to the symmetric distance between country 1 and 2, the price index in country 3 does not depend on the spatial distribution of economic activities within the Union (formed by country 1 and 2).

From (6) and (7), the equilibrium prices and the respective conditions for positive operating profits are:

$$p_{11,t} = \frac{2[a + MC(b + cE)] + cE\,[(1 - x_t)\tilde{n}T_U + (1 - \tilde{n})T_F]}{2(2b + cE)} > MC \tag{8}$$

$$p_{22,t} = \frac{2[a + MC(b + cE)] + cE\,[x_t\tilde{n}T_U + (1 - \tilde{n})T_F]}{2(2b + cE)} > MC$$

$$p_{33,t} = \frac{2[a + MC(b + cE)] + cE\tilde{n}T_F}{2(2b + cE)} > MC$$

$$p_{rs,t} = p_{ss,t} + \frac{T_{rs}}{2} > MC + T_{rs} \qquad \text{for} \qquad r \neq s.$$

Taken together, the constraints reduce to the following.

$$p_{11,t} > MC + \frac{T_F}{2} > MC + \frac{T_U}{2}$$

$$p_{22,t} > MC + \frac{T_F}{2} > MC + \frac{T_U}{2}$$

$$p_{33,t} > MC + \frac{T_F}{2} \tag{9}$$

For simplicity and because we are mainly interested in the fixed point properties, we assume that the parameters satisfy these conditions. However, they deserve further investigation because they may also provide insights into the "zero trade" puzzle (see Trefler 1995, and Anderson and van Wincoop 2004). In the present study, we do not address this issue.

Note that prices are no longer a fixed mark-up on marginal costs, as found in the models with isoelastic demand functions. Two more specific observations are appropriate, as follows. First, all of the equilibrium prices depend on trade costs, not only export prices (i.e., p_{rs} with $r \neq s$) that directly involve trade costs, but also prices for sales within the country (i.e., for p_{rs} with $r = s$). This is because of competition from producers located in other countries who sell in the country under consideration, which is weaker when the trade costs are higher (in this sense, trade costs shelter from competition). For example, if we take p_{11}: it increases if there is less local competition, i.e., if fewer firms are located in country 1 and more firms are located in country 2 (as indicated by a higher $(1 - x_t)\tilde{n}E$) and in country 3 (as indicated by a higher $(1 - \tilde{n})E$), and if local firms are protected by higher transport costs. Firm migration, which is reflected by a change in x_t, modifies the distribution of competitors, thereby introducing an additional competition effect and the prices change accordingly (which would not be the case with isoelastic demand functions). Analogous arguments apply for p_{22} and p_{33}.

Second, prices in one market are no longer equal because entrepreneurs from other countries set prices higher than the incumbent firms, the difference being half of the trade costs (a result that is reminiscent of the standard analysis of monopoly price setting with linear demand functions).

Using demand functions, prices, and price indices, it follows that quantities are given as:

$$q_{rs,t} = (b + cE)(p_{rs,t} - MC - T_{rs}).$$

Note that the conditions specified in Eq. (9) for positive prices and positive operating profits also ensure positive quantities.

3.3 Long-Run Equilibrium

In the long run, entrepreneurs are allowed to move between countries 1 and 2, which they do following differences in the indirect utility. Indirect utility also depends on profit incomes. In order to calculate these profits, the number of customers $C_{r,t}$ (workers and entrepreneurs) residing in country r has to be considered

$$C_{1,t} = \theta L + x_t\tilde{n}E \quad C_{2,t} = \theta L + (1 - x_t)\tilde{n}E \quad C_3 = (1 - 2\theta)L + (1 - \tilde{n})E,$$

and the equilibrium short-run profits $\pi_{rs,t}$ for an entrepreneur located in country r $(= 1, 2)$ and selling to country s $(= 1, 2, 3)$ are:

$$\begin{aligned}
\pi_{11,t} &= (p_{11,t} - MC)q_{11,t}C_{1,t} = (p_{11,t} - MC)^2(b + cE)C_{1,t} \\
\pi_{12,t} &= (p_{12,t} - MC - T_U)q_{12,t}C_{2,t} = (p_{12,t} - MC - T_U)^2(b + cE)C_{2,t} \\
\pi_{13,t} &= (p_{13,t} - MC - T_F)q_{13,t}C_3 = (p_{13,t} - MC - T_F)^2(b + cE)C_3 \\
\pi_{21,t} &= (p_{21,t} - MC - T_U)q_{21,t}C_{1,t} = (p_{21,t} - MC - T_U)^2(b + cE)C_{1,t} \\
\pi_{22,t} &= (p_{22,t} - MC)q_{22,t}C_{2,t} = (p_{22,t} - MC)^2(b + cE)C_{2,t} \\
\pi_{23,t} &= (p_{23,t} - MC - T_F)q_{23,t}C_3 = (p_{23,t} - MC - T_F)^2(b + cE)C_3.
\end{aligned}$$

Note that by symmetry, $\pi_{13,t} = \pi_{23,t}$. The total profit income for an entrepreneur residing in country r $(= 1, 2)$ is:

$$\pi_{r,t} = \pi_{r1,t} + \pi_{r2,t} + \pi_{r3,t}.$$

The consumer surplus in country r $(= 1, 2)$ is:

$$S_{r,t} = \frac{a^2E}{2b} - aP_{r,t} - \frac{c}{2}P_{r,t}^2 + \\ + \frac{b + cE}{2}E\left[x_t\tilde{n}\left(p_{1r,t}\right)^2 + (1 - x_t)\,\tilde{n}\left(p_{2r,t}\right)^2 + (1 - \tilde{n})\left(p_{3r,t}\right)^2\right].$$

Finally, the indirect utilities for entrepreneurs located in country r $(= 1, 2)$ are as follows.

$$V_{r,t} = S_{r,t} + \pi_{r,t} + \overline{C}_A$$

In the long run, the entrepreneurs within the Union move to the country with the highest indirect utility, so the local share of entrepreneurs will increase. The state variable of the full dynamic system is x_t, with $0 \le x_t \le 1$, and we specify a mobility hypothesis that is reminiscent of the replicator dynamics used widely in evolutionary game theory:

$$x_{t+1} = \begin{cases} 0 & if \quad K(x_t) < 0 \\ K(x_t) & if \ 0 \le K(x_t) \le 1 \\ 1 & if \quad K(x_t) > 1, \end{cases}$$

where

$$K(x_t) = x_t\left[1 + \gamma\frac{V_{1,t} - [x_tV_{1,t} + (1 - x_t)V_{2,t}]}{x_tV_{1,t} + (1 - x_t)V_{2,t}}\right] \\ = x_t\left[1 + \gamma(1 - x_t)\frac{V_{1,t} - V_{2,t}}{x_tV_{1,t} + (1 - x_t)V_{2,t}}\right], \tag{10}$$

and where γ is a parameter that indicates how quickly entrepreneurs react to differences in the indirect utility. Note that the indirect utilities can be expressed as functions only of x_t, with $0 \le x_t \le 1$, and the above system of equations defines a one-dimensional piecewise smooth nonlinear map. The difference in indirect utilities can be reformulated as

$$V_1(x_t) - V_2(x_t) = G(\cdot)\,(x_t - 0.5)\,T_U\left(T_U^{crit} - T_U\right)$$

with

$$G(\cdot) = \frac{\tilde{n}E\,(b + cE)}{4\,(2b + cE)^2}\left(E^2c^2\,(3 - \tilde{n}) + 12b^2 + 12Ebc + 8\theta Lbc + 4E\theta Lc^2\right)$$

and

$$T_U^{crit} = \frac{2cE(1-\tilde{n})(4b+3cE)T_F + 8(3b+2cE)(a-bMC)}{4\theta Lc(2b+cE) + [c^2(3-\tilde{n})E^2 + 12b(b+cE)}.$$

Therefore,

$$K(x_t) = x_t \left[1 + \gamma(1-x_t) \frac{G(\cdot)\,(x_t-0.5)\,T_U\left(T_U^{crit}-T_U\right)}{x_t V_{1,t} + (1-x_t)V_{2,t}} \right].$$

This map has three fixed points: the symmetric equilibrium $x^{sym} = 0.5$, where economic activity is spread evenly between the two countries in the Union, and two core periphery equilibria $x^{CP0} = 0$ and $x^{CP1} = 1$, where economic activity is agglomerated in one of the two Union countries.

In the following, we study the properties of these fixed points.

4 Dynamic Analysis and Fixed Point Properties

4.1 Local Stability Properties

Using the well-known conditions on the derivatives, the symmetric equilibrium is (locally) stable for

$$-1 < \left. \frac{\partial K(x_t)}{\partial x_t} \right|_{x_t=\frac{1}{2}} < 1. \tag{11}$$

The right inequality holds provided that

$$T_U > T_U^{crit}.$$

At $T_U = T_U^{crit}$, the map is linear and the symmetric fixed point loses stability through a degenerated pitchfork bifurcation, where all allocations $0 \leq x_t \leq 1$ are stable (although not asymptotically stable). For $T_U < T_U^{crit}$, the two core periphery equilibria are the only attracting fixed points. This corresponds to the typical bifurcation scenario found in many NEG models, where T_U^{crit} represents the so-called break point value of T_U. The only difference is that in our model, we do not find a parameter range for which attracting symmetric and attracting core periphery equilibria coexist, and the transition is abrupt. This is due to the degenerated nature of the pitchfork bifurcation (for details, see Sushko and Gardini (2010)).

4.2 *Properties of the Symmetric Fixed Point*

In this section, we assume that the left inequality in (11) holds (we return to this question in the next section) and we analyze the properties of the symmetric equilibrium, where the mobile firms are distributed equally between the two countries in the Union, $x_t = 0.5$. The allocation of firms does not change further in a steady state, so we can focus on one (representative) firm. In particular, we are interested in the effects of a deeper integration of the Union, which is represented by a reduction in T_U, on the following economic variables:

- Total output of manufactured goods: the total output corresponds to the employment of unqualified labor in the manufacturing sector. The regional endowment with unqualified labor is fixed, so any change in manufacturing employment is reflected by an opposite change in agricultural employment. Thus, we interpret a change in the total manufacturing output as indicating a change in sectoral specialization.
- Total exports: we use the total exports to analyze trade creation effects.
- Exports within the Union and exports to the outside country: we use this to analyze trade diversion effects.

In the symmetric equilibrium, the total output of one firm in country 1 (selling to all three countries) $q_1^s = q_{11}^s + q_{12}^s + q_{13}^s$ is given as (country 2 is symmetric):

$$q_1^s = q_2^s = \frac{(E+L)(\alpha - MC)}{2(\beta-\delta)+\delta E} - \frac{[2(\beta-\delta)+\delta E(1-\tilde{n})]\,C_1^s}{2(\beta-\delta)[2(\beta-\delta)+\delta E]}T_U - \frac{2(\beta-\delta)C_3+\delta E(1-\tilde{n})\,\Delta C}{2(\beta-\delta)[2(\beta-\delta)+\delta E]}T_F,$$

where the superscript s indicates values for the symmetric equilibrium, and where we substituted back the original parameters α, β, and γ. $C_1^s = C_2^s = \theta L + 0.5\tilde{n}E$ and $C_3 = L+E-C_1^s-C_2^s = (1-2\theta)L+(1-\tilde{n})E$ denote the number of customers residing in each of the countries (and thus the market size); and $\Delta C = C_3 - C_1^s - C_2^s$ is the difference in market size. When $\Delta C > 0$, the market inside the Union is smaller than the market outside the region. Note that C_1^s and C_2^s depend on the allocation of the mobile factor, whereas this is not the case for C_3 and ΔC.

For $\delta = 0$, this expression simplifies to

$$q_1^s = q_2^s = \frac{(E+L)(\alpha - MC)}{2\beta} - \frac{C_1^s}{2\beta}T_U - \frac{\Delta C}{2\beta}T_F.$$

A reduction in T_U increases manufacturing output within the Union and this effect is stronger when the market inside the Union is bigger (higher C_1^s). Both results also hold in the general case, while the cross-derivatives with respect to the other parameters (particularly of E, $\tilde{n}$, and δ) are ambiguous.

The total output of one firm in country 3 (selling to all three countries) $q_3^s = q_{31}^s + q_{32}^s + q_{33}^s$ is given as follows.

$$q_3^s = \frac{(E+L)(\alpha - MC)}{2(\beta-\delta)+\delta E} + \frac{\delta E \tilde{n} C_1^s}{2(\beta-\delta)[2(\beta-\delta)+\delta E]} T_U -$$

$$-\frac{4(\beta-\delta+\delta E\tilde{n}) C_1^s - \delta E \tilde{n}(E+L)}{2(\beta-\delta)[2(\beta-\delta)+\delta E]} T_F$$

For $\delta = 0$, this expression simplifies to

$$q_3^s = \frac{(E+L)(\alpha - MC)}{2\beta} - \frac{C_1^s}{\beta} T_F.$$

It is interesting to note that in the special case where $\delta = 0$ (which represents a high preference for product variety), the output in country 3 is independent of T_U. In the general case where $0 < \delta < \beta$, a reduction in the trade costs T_U decreases the manufacturing output in country 3. It can be shown that this effect is stronger when δ is higher (a lower preference for product variety), when θL is higher (which leads to a bigger market within the Union), and when E and $\tilde{n}$ are higher (which also lead to a bigger market within the Union, as well as stronger competition at the same time).

Thus, deeper integration within the Union, as reflected by a reduction in the internal trade costs T_U, leads to increased manufacturing output and employment in both countries in the Union (and to a corresponding reduction in agricultural output), i.e., the Union specializes in manufacturing, whereas the manufacturing output and employment are reduced in the outside country 3 (and the agricultural sector expands correspondingly), i.e., country 3 specializes in agriculture.

Next, we consider the analysis of international trade. The exports by one firm located within the Union selling to the other country in the Union are given as follows.

$$q_{12}^s = q_{21}^s = \frac{C_1^s(\alpha - MC)}{2(\beta-\delta)+\delta E} - \frac{4(\beta-\delta)+\delta E(2-\tilde{n})}{(2(\beta-\delta)+\delta E)\,4(\beta-\delta)} C_1^s T_U +$$

$$+\frac{\delta E(1-\tilde{n})}{(2(\beta-\delta)+\delta E)\,2(\beta-\delta)} C_1^s T_F$$

For $\delta = 0$, this expression simplifies to

$$q_{12}^s = q_{21}^s = \frac{C_1^s(\alpha - MC)}{2\beta} - \frac{C_1^s}{2\beta} T_U.$$

A reduction in T_U leads to increased exports within the Union. This effect is stronger when θL is higher, which reflects a bigger market within the Union, while the cross-derivatives with respect to $\delta, E, \tilde{n}$ are ambiguous.

The exports by one firm located within the Union to the outside country are:

$$q^s_{13} = q^s_{23} = \frac{C_3(\alpha - MC)}{2(\beta - \delta) + \delta E} + \frac{2(\beta - \delta) - \delta E(1 - \tilde{n})}{2(\beta - \delta)[2(\beta - \delta) + \delta E]} C_3 T_F.$$

For $\delta = 0$, this expression simplifies to

$$q^s_{13} = q^s_{23} = \frac{C_3(\alpha - MC)}{2\beta} + \frac{C_3}{2\beta} T_F.$$

It is interesting to note that the Union's exports to the outside region are not affected by a change in the internal trade costs T_U.

Thus, with deeper integration inside the Union, which is reflected by a reduction in T_U, the total exports by the Union increase (because international trade within the Union increases whereas it remains unchanged with the outside country).

Now, we consider a firm located in country 3 outside the Union, where its exports to a country within the Union (to country 1 or 2) are as follows.

$$q^s_{31} = q^s_{32} = \frac{C^s_1(\alpha - MC)}{2(\beta - \delta) + \delta E} + \frac{\delta E \tilde{n}}{4(\beta - \delta)[2(\beta - \delta) + \delta E]} C^s_1 T_U -$$

$$-\frac{2(\beta - \delta) + \delta E \tilde{n}}{2(\beta - \delta)[2(\beta - \delta) + \delta E]} C^s_1 T_F$$

For $\delta = 0$, this expression simplifies to

$$q^s_{31} = q^s_{32} = \frac{C^s_1(\alpha - MC)}{2\beta} - \frac{C^s_1}{2\beta} T_F.$$

In the special case where $\delta = 0$, a reduction in T_U has no effect. In the general case, it reduces the exports by one firm located in country 3 to countries 1 and 2. The effect is stronger when δ is higher (a lower preference for product variety), when θL is higher (which reflects a bigger market within the Union), and when E and $\tilde{n}$ are higher (which also reflects a bigger market within the Union, as well as stronger competition at the same time).

Finally, we analyze how total exports $q^s_{12} + q^s_{21} + q^s_{13} + q^s_{23} + q^s_{31} + q^s_{32} = 2\left(q^s_{12} + q^s_{13} + q^s_{31}\right)$ change with a reduction in the internal trade costs T_U. We have shown that q_{13}^s does not change, q_{12}^s increases, and q_{31}^s decreases; therefore, we focus on the last two terms:

$$q_{12}^s + q_{31}^s = \frac{2C_1^s(\alpha - MC)}{2(\beta-\delta)+\delta E} - \frac{2(\beta-\delta)+\delta E(1-\tilde{n})}{2(\beta-\delta)[2(\beta-\delta)+\delta E]}C_1^s T_U - \\ -\frac{2(\beta-\delta)-\delta E(1-2\tilde{n})}{2(\beta-\delta)[2(\beta-\delta)+\delta E]}C_1^s T_F.$$

For $\delta = 0$, this expression simplifies to

$$q_{12}^s + q_{31}^s = \frac{C_1^s(\alpha - MC)}{2\beta} - \frac{C_1^s T_U}{2\beta} - \frac{C_1^s T_F}{2\beta}.$$

A deeper integration within the Union, which is reflected by a reduction in T_U, increases total exports; thus, the increase in the Union's exports is bigger than the reduction in the exports by the outside country 3. To summarize our results. In the symmetric equilibrium, a deeper integration within the Union leads to a specialization in manufacturing (agriculture) for the Union (outside country), and we also observe trade diversion and trade creation effects.

4.3 *Transition to a Core Periphery Equilibrium*

As shown above, the symmetric fixed point loses stability when the internal trade costs fall below the critical value T_U^{crit}. This transition gives rise to two core periphery equilibria, where industrial activity within the Union is agglomerated in one of the two member countries and the market sizes within the Union change accordingly. If country 1 is the core, the market sizes are $C_1^{CP1} = \theta L + \tilde{n}E$ and $C_2^{CP1} = \theta L$, and the industrial output of one firm located in country 1 is as follows.

$$q_1^{CP1} = \frac{(E+L)(\alpha - MC)}{2(\beta-\delta)+\delta E} - \frac{[2(\beta-\delta)+\delta E(1-\tilde{n})]C_1^{CP1}}{2(\beta-\delta)[2(\beta-\delta)+\delta E]}T_U - \\ -\frac{2(\beta-\delta)C_3+\delta E(1-\tilde{n})\Delta C}{2(\beta-\delta)[2(\beta-\delta)+\delta E]}T_F$$

It is interesting to note that

$$q_1^s - q_1^{CP1} = -\frac{[2(\beta-\delta)+\delta E(1-\tilde{n})]E\tilde{n}}{4(\beta-\delta)[2(\beta-\delta)+\delta E]}T_U < 0. \tag{12}$$

Thus, at T_U^{crit}, reducing T_U leads to the agglomeration of industrial activity in country 1 and to an expansion of the output by each single firm located in the

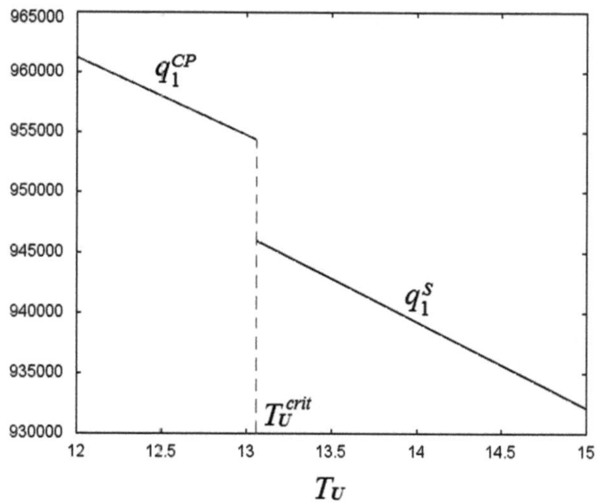

Fig. 1 Bifurcation diagram illustrating how output depends on transportation costs within the Union

agglomeration area. Figure 1 visualizes this analytic result where the bifurcation diagram plots T_U on the abscissa and q_1 on the ordinate. We used the following parameter values: $a = 900$, $b = 2$, $c = 1$, $E = 100$, $\tilde{n} = 0.6$, $L = 1000$, $\theta = 0.3$, $T_L = 20$, and $MC = 0.5$. These parameters imply that $T_U^{crit} \approx 13.054$. We varied the trade costs with the Union in a range that crossed the breakpoint value $12 \leq T_U \leq 15 < T_F$ and we checked that the nonnegativity conditions specified in (9) were satisfied during the simulation. The figure illustrates the increase in the quantity produced if trade costs are reduced within the Union and it clearly shows the jump in output at the breakpoint (which leads to the agglomeration of economic activity).

This result is consistent with economic intuition since agglomeration leads to more competition in country 1 and monopolists react by reducing prices and increasing quantities. We confirm this intuition by taking another look at the prices set by firms in country 1 (see Eq. (8), where we substitute back the original parameters for α, β, and δ):

$$p_{11} = A + \frac{\delta\,(1-x)\,\tilde{n}E}{2\,(\beta-\delta)+\delta E}\frac{T_U}{2} + \frac{\delta E\,(1-\tilde{n})}{2\,(\beta-\delta)+\delta E}\frac{T_F}{2}$$

$$p_{12} = A + \left(1 + \frac{\delta x \tilde{n}E}{2\,(\beta-\delta)+\delta E}\right)\frac{T_U}{2} + \frac{\delta E\,(1-\tilde{n})}{2\,(\beta-\delta)+\delta E}\frac{T_F}{2}$$

$$p_{13} = A + \left(1 + \frac{\delta \tilde{n}E}{2\,(\beta-\delta)+\delta E}\right)\frac{T_F}{2},$$

where $A = \frac{(\beta-\delta)(\alpha+MC)+MC\delta E}{2(\beta-\delta)+\delta E}$ is a parameter cluster that is independent of trade costs and the regional distribution of industrial activity. Note that even p_{11} depends positively upon the trade costs, although firms selling to the market where they are located do not incur trade costs (in contrast to their foreign competitors). This effect is weaker when local competition is stronger. For T_U, the relevant indicator is the number of firms located in country 1 (measured by x), and for T_F, the relevant indicator is the number of firms located within the Union (measured by $\tilde{n}$). The price set by one firm located in country 1 for sales in country 2, p_{12}, involves the direct trade costs $\frac{T_U}{2}$, half of which are passed on to the consumers, as expected for monopoly pricing with linear demand functions. However, the pass-through of the transport costs also depends on competition. When fewer firms are located in country 2 (reflected by a higher x), the the pass-through is higher for T_U. For T_F, a similar argument to that given above applies. The interpretation for p_{13} follows the same lines.

At T_U^{crit}, reducing T_U increases x from 0.5 to 1, while p_{11} is reduced and p_{12} increases, but p_{13} does not change. As shown above (see Eq. (12)), the Union will produce more of the industrial commodities (higher manufacturing employment, i.e., increased specialization), but it will export the same quantities to the third, outside country (as shown by the unchanged p_{13}).

4.4 Further Dynamic Considerations

In this section,[2] we give some further considerations on the model dynamics. In contrast to the findings in the previous sections, the following results are primarily related to the specification in discrete time (and not in continuous time). We prefer this specification because it is more suitable for reflecting delays, which are necessarily involved in factor mobility processes. It is interesting to note that for higher values of T_U (and for higher values of γ), the symmetric fixed point looses stability through a flip bifurcation, which gives rise to a period-2 cycle at a bifurcation value that is defined implicitly as follows.

$$-1 = \left. \frac{\partial K(x_t)}{\partial x_t} \right|_{x_t=\frac{1}{2}}$$

An analytic expression is not available but we provide some simulations to illustrate the implied dynamic behavior.

Figure 2a, which is based on the same parameters as Fig. 1, depicts the full bifurcation scenario with respect to the trade costs, where the abscissa measures T_U and the ordinate now refers to x_t, i.e., to the share of mobile capital located in country 1. Crossing T_U^{crit} from below, the transition from a core periphery to a symmetric equilibrium discussed in the previous section is clearly visible. Further increasing the trade costs leads to a period doubling bifurcation and one period-2 cycle, which

[2]Ingrid Kubin dedicates this section to Karl Framer with whom she detected the complexity of complex dynamics almost 30 years ago.

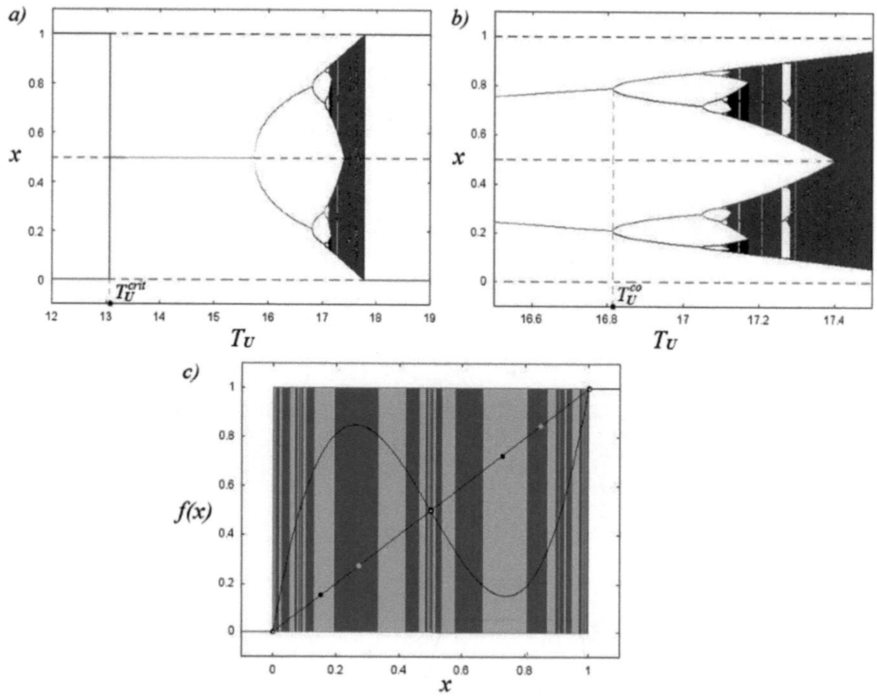

Fig. 2 One-dimensional bifurcation diagram (2a), enlargement (2b), and underlying map (2c) illustrating the full dynamics of the capital share in country 1, which depend on the trade costs within the Union

then undergoes a pitchfork bifurcation. The enlargement of the bifurcation diagram in Fig. 2b shows that two period-2 attractors coexist for $T_U > T_U^{co}$ (one is depicted in grey and the other in black). Each of these two cycles undergoes a cascade of period doubling bifurcations, followed by a "logistic" bifurcation scenario involving cycles of high periodicity and complex and chaotic attractors. Finally, as shown in Fig. 2a, the long-run behavior of the model is again a core periphery equilibrium.

As noted above, due to the degenerate nature of the pitchfork bifurcation, the transition from a core periphery to a symmetric equilibrium at T_U^{crit} is abrupt and it does not involve any "overlapping," so for this parameter range, we do not observe the coexistence of any attractors. However, we have just shown that after the pitchfork bifurcation at $T_U = T_U^{co}$, the model involves many instances of coexistence if we allow for cyclical and complex attractors. Figure 2c represents the map where the circles on the 45-degree line correspond to the two coexisting period-2 cycles and the light and dark grey strips indicate the respective basins of attraction, i.e., initial conditions for x_t starting from which gives rise to a time path that ultimately settles on one of the two possible period-2 attractors. The structure of these basins is highly complex, where small shocks to the initial distribution of economic activity (represented in x_t) give rise to different long-term patterns of economic activity.

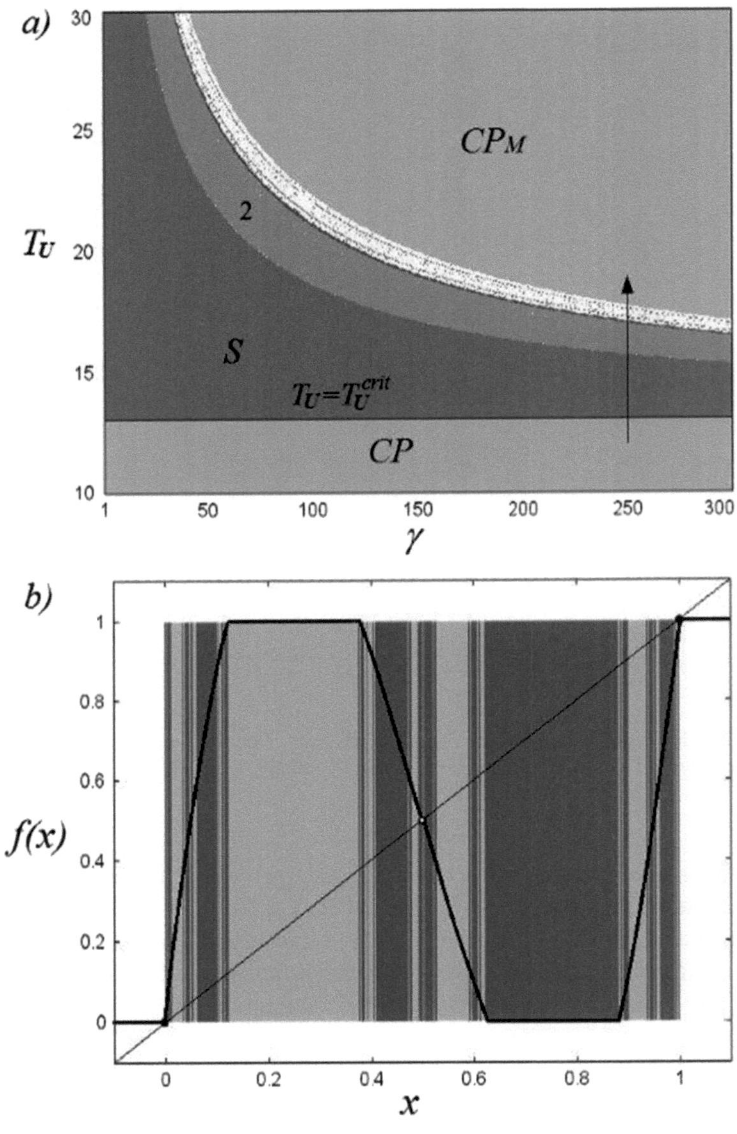

Fig. 3 Two-dimensional bifurcation diagram (3a) and underlying map (3c) illustrating the dynamics of the capital share in country 1, which depend on the two dimensions of integration, i.e., the trade costs within the Union and capital mobility

History matters, and the complex structure of the basins strongly highlights this core theme of the NEG.

Figure 3, which is again based on the same parameter values as those used in Fig. 1, is a two-dimensional bifurcation diagram that represents the dynamic long-run outcome if T_U is varied as well as γ. The parameter combinations in the *CP*

$(S, 2)$ area are connected to the core periphery (symmetric, period-2) attractors in the long-run. The whitish area shows the parameter combinations that lead to attractors with a higher periodicity or to chaotic attractors. The black arrow indicates the parameter variation depicted in the one-dimensional bifurcation diagrams in Fig. 2. The area CP_M deserves one final comment because these parameter combinations ultimately lead to the coexistence of both core periphery equilibria; however, the respective basins of attractions (depicted in Fig. 3b, which is interpreted as Fig. 2c) are highly complex, and the attractors are of the Milnor type (in the sense of Milnor 1985). Even for initial points close to one CP equilibrium, it cannot be stated a priori to which of the two CP fixed points the system will converge (see Commendatore et al. 2014, for a more detailed discussion). Again, history matters and a small perturbation may significantly alter the long-run regional distribution of industrial activity.

5 Conclusions

In this study, we analyzed the effects of trade agreements using a NEG model. In contrast to the standard Heckscher-Ohlin framework, the countries are similar, production involves fixed costs, and thus markets are monopolistically competitive, and some of the productive factors are internationally mobile. We used a footloose entrepreneur model because it allows the mobility of highly qualified entrepreneurial labor, whereas unqualified labor is immobile, which roughly corresponds to stylized facts for the European Union. In addition, we used the linear version of this model because of its higher analytical tractability. We set up the model for a three-country context, where two countries form a Union and the other country is outside.

We studied the effects of deeper integration within the Union primarily with respect to commodity trade, which is represented by a reduction in the internal trade costs in the model. However, we noted that deeper integration within the Union can also reflect a reduction in barriers to factor mobility, and thus we also studied some effects of this type of integration.

We showed that reducing internal trade costs increases the production of manufactured goods in the Union, i.e., the Union specializes in this sector, increases the total exports of the Union, and increases the trade within the Union, i.e., integration leads to specialization, trade creation, and trade diversion. These effects are similar to a Heckscher-Ohlin perspective. However, if the trade costs fall below the so-called breakpoint value, industrial activity is no longer distributed symmetrically between the two countries within the Union, but instead it agglomerates in one of the countries. This effect is genuine to a NEG perspective because it is based on factor mobility. Competition increases in the country where economic activity is agglomerated, and monopolistically competitive firms react by reducing the price and further increasing the manufacturing output, which adds another dimension to the specialization effect of trade integration. Trade with the outside country remains the same, but internal trade increase, and thus trade is created and diverted. Therefore, our analysis of trade agreements from a NEG perspective corroborates and extends the standard Heckscher-Ohlin analysis by showing that deeper integration may lead to agglomeration within the Union. Thus, our initial observation that regional disparities remain stubbornly high within the EU even after an impressive integration process over the last 50 years is not so surprising after all. Finally, we provided more insights into the dynamics involved, where we showed that cyclical and complex patterns are possible in the regional distribution of industrial activities, and that the coexistence of attractors (where the basins of attractions have a complex structure) is far more widespread than suggested by standard analyses of the NEG.

References

Anderson JE, van Wincoop E (2004) Trade costs. J Econ Lit 42(3):691–751

Behrens K (2001) International integration and regional inequalities: how important is national infrastructure? Manch Sch 79(5):952–971

Brülhart M (2011) The spatial effects of trade openness: a survey. Rev World Econ 147(1):59–83

Brülhart M, Crozet M, Koenig P (2004) Enlargement and the EU periphery: the impact of changing market potential. World Econ 27(6):853–875

Commendatore P, Kubin I, Petraglia C, Sushko I (2014) Regional integration, international liberalisation and the dynamics of industrial agglomeration. J Econ Dyn Control 48:265–287

Commendatore P, Filosofo V, Kubin I, Grafeneder-Weissteiner T (2015) Towards a multiregional NEG framework: comparing alternative modelling strategies. In: Commendatore P, Kubin I, Suna Kayam S (eds) Complexity and geographical economics. Dynamic Modeling and Econometrics in Economics and Finance, vol 19. Springer, Cham Heidelberg/New York, Dordrecht/London, pp 13–50

Dorrucci E, Firpo S, Fratzscher M, Monelli FP (2002) European integration: what lessons for other regions? The case of Latin America. ECB working paper 185, 2002

Gáková Z, Dijkstra L (2008) Labour mobility between the regions of the EU-27 and a comparison with the USA. Technical report 2, European Union

Krugman P (1991) Increasing returns and economic geography. J Polit Econ 99(3):483–99

Krugman P, Livas Elizondo P (1996) Trade policy and the third world metropolis. J Dev Econ 49(1):137–150

Milnor J (1985) On the concept of attractor. Commun Math Phys 99:177–195

Obstfeld M, Peri G (1998) Regional non-adjustment and fiscal policy. Econ Policy 13(26):205–259

Ottaviano G, Tabuchi T, Thisse J-F (2002) Agglomeration and trade revisited. Int Econ Rev 43(2):409–435

Paluzie E (2001) Trade policy and regional inequalities. Pap Reg Sci 80(1):67–85

Puga D (2001) European regional policies in light of recent location theories. J Econ Geogr 2(4):373–406

Siebert H (1997) Labor market rigidities: at the root of unemployment in Europe. J Econ Perspect 11(3):37–54

Sushko I, Gardini L (2010) Degenerate bifurcations and border collisions in piecewise smooth 1D and 2D maps. Int J Bifurcation Chaos 20(7):2045–2070

Trefler D (1995) The case of the missing trade and other mysteries. Am Econ Rev 85(5):1029–1046

Wang J, Zheng X-P (2013a) Industrial agglomeration and dispersion in gate and hinterland regions. Ritsumeikan Econ Rev 62(1):39–60

Wang J, Zheng X-P (2013b) Industrial agglomeration: asymmetry of regions and trade costs. Rev Urban Reg Dev Stud 25(2):61–78

Strategic Macroeconomic Policies in a Monetary Union

R. Neck and D. Blueschke

1 Introduction

The economic situation in the European Monetary Union is relatively unstable nowadays due to the economic crisis of 2008–2010 (the Great Recession) and a wide range of structural problems in the affected countries. At the beginning of this crisis policy makers tried to cooperate and to use coordinated anti-cyclical fiscal and monetary policies to reduce the negative impacts of the crisis, giving high importance to GDP growth and reducing unemployment. Unfortunately, the public debt situation worsened dramatically and we have been facing a sovereign debt crisis in Europe since 2010. During this latest crisis politicians have shown no consensus on the best way out of the crisis. And suddenly the European Monetary Union no longer seems to be a union of cooperating partners speaking with one voice but a pool of independent players seeking the profits for their own country only. Strategic considerations play an important role in this situation. Hence, it is useful to run a study of a monetary union using concepts of dynamic game theory.

The framework of dynamic games is suitable for describing the dynamics of a monetary union because a monetary union consists of several players with independent or rather different aims and instruments. Even if there are common, union-wide indexes, each of the players may give different importance (weights) to these targets. In addition, the willingness to cooperate in order to achieve the common goal may be country-specific as well. For these reasons it is indispensable to model the conflicts ('non-cooperation') between the players. Such problems can be best modeled using the concepts and methods of dynamic game theory, which was mostly developed by engineers and mathematicians but which has proved to be

R. Neck (✉) • D. Blueschke
Department of Economics, Alpen-Adria Universität Klagenfurt, Universitätsstrasse 65-67, 9020 Klagenfurt, Austria
e-mail: reinhard.neck@aau.at; dmitri.blueschke@aau.at

B. Bednar-Friedl, J. Kleinert (eds.), *Dynamic Approaches to Global Economic Challenges*, DOI 10.1007/978-3-319-23324-6_4

a valuable analytical tool for economists, too (see, e.g., Basar and Olsder 1999; van Aarle et al. 2002).

In this paper we present an application of the dynamic tracking game framework to a monetary union model. Dynamic games have been used for modeling conflicts between monetary and fiscal policies by several authors. There is also a large body of literature on dynamic conflicts between policy makers from different countries on issues of international stabilization. Both types of conflict are present in a monetary union because a supranational central bank interacts strategically with sovereign governments as national fiscal policy makers in the member states. Such conflicts can be analyzed using either large empirical macroeconomic models (e.g. Haber et al. 2002) or small stylized models (e.g. Dixit and Lambertini 2001; Neck and Behrens 2009; Neck and Blueschke 2014). We follow the latter line of research and use a small stylized nonlinear two-country macroeconomic model of a monetary union (called MUMOD1) for analysing the interactions between fiscal (governments) and monetary (common central bank) policy makers, assuming different objective functions of these decision makers. Using the OPTGAME algorithm we calculate solutions for two game strategies: one cooperative (Pareto optimal) and one non-cooperative game type (the Nash game for the feedback information pattern).

Applying the OPTGAME algorithm to the MUMOD1 model we show how the policy makers react optimally to demand shocks. Applying a demand shock we reproduce the basic dynamics of the economic crisis 2008–2010 and the sovereign debt crisis in Europe. Furthermore we assume several future demand side shocks acting on the monetary union and analyze what would be the best strategies in such a hypothetical case.

Dropping demand is a very important problem in the EMU nowadays. In contrast, high inflation is not a problem in the EMU at the moment. In June 2014 the ECB decided to lower the interest rates, which resulted in a prime rate of 0.15 % (in the meantime it has been further decreased to 0.05 %) and a deposit facility rate of −0.10 % (a negative interest rate for the first time in the history of the ECB). This step was taken "since euro area inflation is expected to remain considerably below 2 % for a prolonged period".[1] Therefore, in this study we do not model the situation of a future supply side shock, such as an oil price shock, which would increase inflation.

2 Nonlinear Dynamic Tracking Games

We consider nonlinear dynamic game-theoretic problems which are given in tracking form. The players aim at minimizing quadratic deviations of the equilibrium values from given target (desired) values. Thus each player minimizes an objective

[1] See the official press release of the ECB from 5th June 2014.

function J^i given by:

$$\min_{u_1^i,\ldots,u_T^i} J^i = \sum_{t=1}^{T} L_t^i(x_t, u_t^1, \ldots, u_t^N), \qquad i = 1, \ldots, N, \tag{1}$$

with

$$L_t^i(x_t, u_t^1, \ldots, u_t^N) = \frac{1}{2}[X_t - \tilde{X}_t^i]'\Omega_t^i[X_t - \tilde{X}_t^i], \qquad i = 1, \ldots, N. \tag{2}$$

The parameter N denotes the number of players (decision makers). T is the terminal period of the finite planing horizon, i.e. the duration of the game. X_t is an aggregated vector

$$X_t := [x_t \;\; u_t^1 \;\; u_t^2 \;\; \ldots \;\; u_t^N]', \tag{3}$$

which consists of an $(n_x \times 1)$ vector of state variables

$$x_t := [x_t^1 \;\; x_t^2 \;\; \ldots \;\; x_t^{n_x}]' \tag{4}$$

and N $(n_i \times 1)$ vectors of control variables determined by the players $i = 1, \ldots, N$:

$$\begin{aligned} u_t^1 &:= [u_t^{11} \;\; u_t^{12} \;\; \ldots \;\; u_t^{1n_1}]', \\ u_t^2 &:= [u_t^{21} \;\; u_t^{22} \;\; \ldots \;\; u_t^{2n_2}]', \\ &\vdots \\ u_t^N &:= [u_t^{N1} \;\; u_t^{N2} \;\; \ldots \;\; u_t^{Nn_N}]'. \end{aligned} \tag{5}$$

Thus X_t (for all $t = 1, \ldots, T$) is an r-dimensional vector, where

$$r := n_x + n_1 + n_2 + \cdots + n_N. \tag{6}$$

The desired levels of the state variables and the control variables of each player enter the quadratic objective functions (as given by Eqs. (1) and (2)) via the terms

$$\tilde{X}_t^i := [\tilde{x}_t^i \;\; \tilde{u}_t^{i1} \;\; \tilde{u}_t^{i2} \;\; \ldots \;\; \tilde{u}_t^{iN}]'. \tag{7}$$

Finally, Eq. (2) contains an $(r \times r)$ penalty matrix Ω_t^i $(i = 1, \ldots, N)$, weighting the deviations of states and controls from their desired levels in any time period t $(t = 1, \ldots, T)$. Thus the matrices

$$\Omega_t^i = \begin{bmatrix} Q_t^i & 0 & \cdots & 0 \\ 0 & R_t^{i1} & 0 & \vdots \\ \vdots & 0 & \ddots & 0 \\ 0 & \cdots & 0 & R_t^{iN} \end{bmatrix}, \; i = 1, \ldots, N, \; t = 1, \ldots, T, \tag{8}$$

are of block-diagonal form, where the blocks Q_t^i and R_t^{ij} $(i,j = 1,\ldots,N)$ are symmetric. These blocks Q_t^i and R_t^{ij} correspond to penalty matrices for the states and the controls, respectively. The matrices $Q_t^i \geq 0$ are positive semi-definite for all $i = 1,\ldots,N$; the matrices R_t^{ij} are positive semi-definite for $i \neq j$ but positive definite for $i = j$. This guarantees that the matrices $R_t^{ii} > 0$ are invertible, a necessary prerequisite for the analytical tractability of the algorithm.

In a frequent special case, a discount factor α is used to calculate the penalty matrix Ω_t^i in time period t:

$$\Omega_t^i = \alpha^{t-1}\Omega_0^i, \tag{9}$$

where the initial penalty matrix Ω_0^i of player i is given.

The dynamic system, which constrains the choices of the decision makers, is given in state-space form by a first-order system of nonlinear difference equations:

$$x_t = f(x_{t-1}, x_t, u_t^1, \ldots, u_t^N, z_t), \quad x_0 = \bar{x}_0. \tag{10}$$

$\bar{x}_0$ contains the initial values of the state variables. The vector z_t contains non-controlled exogenous variables. f is a vector-valued function where f^k $(k = 1,\ldots,n_x)$ denotes the kth component of f. For the algorithm, we require that the first and second derivatives of the system function f with respect to x_t, x_{t-1} and $u_t^1,\ldots,u_t^N$ exist and are continuous. The assumption of a first-order system of difference equations as stated in (10) is not really restrictive as higher-order difference equations can be reduced to systems of first-order difference equations by suitably redefining variables as new state variables and augmenting the state vector.

Equations (1), (2) and (10) define a nonlinear dynamic tracking game problem to be solved. That means, we try to find N trajectories of control variables u_t^i, $i = 1,\ldots,N$, which minimize the postulated objective functions subject to the dynamic system. Using the OPTGAME algorithm (see Blueschke et al. 2013) we are able to solve the stated dynamic tracking game problem. Applying the OPTGAME algorithm to the MUMOD1 model, we calculate optimal macroeconomic policies in a monetary union for different solution concepts, namely the noncooperative feedback Nash equilibrium solution, which is subgame perfect, and an efficient Pareto optimal solution, which is not an equilibrium of the dynamic game but requires some external mechanism of implementation to become effective. As usual in dynamic game theory (cf. Basar and Olsder 1999), the dynamic system, the exogenous variables and the objective functions of the players are assumed to be common knowledge, and the players know the state vector of the last period when deciding on their control variables in the current period.

3 The MUMOD1 Model

In this paper we use a simplified model of a monetary union which is called MUMOD1 and which slightly improves on the one introduced in Blueschke and Neck (2011) in order to derive optimal fiscal and monetary policies of the economies in a monetary union. The model is calibrated so as to deal with the problem of public debt targeting in a situation that resembles the one currently prevailing in the European Union, but no attempt is made to describe a monetary union in general or the EMU in every detail. The model builds on discrete data, which is a popular method in economics. One of the most important features of our model is the fact that it allows for different kinds of exogenous shocks acting in a symmetric or an asymmetric way on the economies in the monetary union. Analyzing the impacts of these different shocks allows us to gain basic insights into the dynamics of a monetary union.

It should be stressed that the MUMOD1 model is strictly confined to be used for analyzing short run effects of fiscal and monetary policies. For this reason, it is formulated in terms of deviations from an (unspecified) long run steady state growth path of the economies under consideration. Hence we separate long run growth, which may follow a balanced path according to a neoclassical growth model, from short and medium run developments which are the exclusive concern of this paper. Long run effects of public debt consolidation as analyzed frequently in the literature (e.g. Farmer 2006, 2011; Farmer and Zotti 2010a,b; Farmer and Schelnast 2013) cannot be dealt with using our model. Intertemporal effects arising in the goods market, from inflationary expectations and from the government budget constraint are assumed to be short (or medium) lived and without any effects on long run values of the model variables.

In this paper we introduce two sequences of different demand side shocks on the monetary union. The first sequence includes a negative asymmetric demand side shock aimed at describing the dynamics in a monetary union in a situation similar to the economic crisis (2007–2010) and the sovereign debt crisis in Europe (since 2010). The second sequence of shocks (different kinds of demand side shocks) serves to discuss the best macroeconomic policy strategies for possible future scenarios. Before we present these studies it is necessary to describe the model in detail.

In the following, capital letters indicate nominal values, while lower case letters correspond to real values. Variables are denoted by Roman letters and model parameters are denoted by Greek letters. Three active policy makers are considered: the governments of the two countries responsible for decisions about fiscal policy and the common central bank of the monetary union controlling monetary policy. The two countries are labeled 1 and 2 or core and periphery respectively. The idea is to create a stylized model of a monetary union consisting of two homogeneous blocs of countries, which in the current European context might be identified with the stability-oriented, core bloc and the periphery bloc (countries with problems due to high public debt and/or high budget deficits).

The model is formulated in terms of deviations from a long-run growth path. The goods markets are modeled for each country by a short-run income-expenditure equilibrium relation (IS or AD curve). The two countries under consideration are linked through their goods markets, namely exports and imports of goods and services. The common central bank decides on the prime rate, that is, a nominal rate of interest under its direct control (for instance, the rate at which it lends money to private banks).

Real output (or the deviation of short-run output from a long-run growth path) in country i ($i = 1, 2$) at time t ($t = 1, \ldots, T$) is determined by a reduced form demand-side equilibrium equation:

$$y_{it} = \delta_i(\pi_{jt} - \pi_{it}) - \gamma_i(r_{it} - \theta) + \rho_i y_{jt} - \beta_i \pi_{it} + \kappa_i y_{i(t-1)} - \eta_i g_{it} + zd_{it}, \qquad (11)$$

for $i \neq j$ ($i,j = 1, 2$). The variable π_{it} denotes the rate of inflation in country i, r_{it} represents country i's real rate of interest and g_{it} denotes country i's real fiscal surplus (or, if negative, its fiscal deficit), measured in relation to real GDP. g_{it} in (11) is assumed to be country i's fiscal policy instrument or control variable. The natural real rate of output growth, $\theta \in [0, 1]$, is assumed to be equal to the natural real rate of interest. The parameters $\delta_i, \gamma_i, \rho_i, \beta_i, \kappa_i, \eta_i$, in (11) are assumed to be positive. The variables zd_{1t} and zd_{2t} are non-controlled exogenous variables and represent exogenous demand side shocks in the goods market.

For $t = 1, \ldots, T$, the current real rate of interest for country i ($i = 1, 2$) is given by:

$$r_{it} = I_{it} - \pi^e_{it}, \qquad (12)$$

where π^e_{it} denotes the expected rate of inflation in country i and I_{it} denotes the nominal interest rate for country i, which is given by:

$$I_{it} = R_{Et} - \lambda_i g_{it} + \chi_i D_{it}, \qquad (13)$$

where R_{Et} denotes the prime rate determined by the central bank of the monetary union (its control variable); $-\lambda_i$ and χ_i (λ_i and χ_i are assumed to be positive) are risk premia for country i's fiscal deficit and public debt level respectively. This allows for different nominal (and a fortiori also real) rates of interest in the union in spite of a common monetary policy due to the possibility of default or similar risk of a country (a bloc of countries) with high government deficit and debt.

The inflation rates for each country $i = 1, 2$ and $t = 1, \ldots, T$ are determined according to an expectations-augmented Phillips curve, i.e. the actual rate of inflation depends positively on the expected rate of inflation and on the goods market excess demand (a demand-pull relation):

$$\pi_{it} = \pi^e_{it} + \xi_i y_{it} + zs_{it}, \qquad (14)$$

where ξ_1 and ξ_2 are positive parameters; zs_{1t} and zs_{2t} denote non-controlled exogenous variables and represent possible exogenous supply side shocks; π^e_{it} denotes the rate of inflation in country i expected to prevail during time period t, which is formed at (the end of) time period $t-1$. Inflationary expectations are formed according to the hypothesis of adaptive expectations:

$$\pi^e_{it} = \varepsilon_i \pi_{i(t-1)} + (1-\varepsilon_i)\pi^e_{i(t-1)}, \tag{15}$$

where $\varepsilon_i \in [0, 1]$ are positive parameters determining the speed of adjustment of expected to actual inflation.

The average values of output and inflation in the monetary union are given by:

$$y_{Et} = \omega y_{1t} + (1-\omega)y_{2t}, \quad \omega \in [0, 1], \tag{16}$$

$$\pi_{Et} = \omega \pi_{1t} + (1-\omega)\pi_{2t}, \quad \omega \in [0, 1]. \tag{17}$$

The parameter ω expresses the weight of country 1 in the economy of the whole monetary union as defined by its output level. The same weight ω is used for calculating union-wide inflation in Eq. (17).

The government budget constraint is given as an equation for government debt of country i ($i = 1, 2$):

$$D_{it} = (1 + BI_{i,t-1} - \pi^e_{i,t-1})D_{i,t-1} - g_{it}, \tag{18}$$

where D_i denotes real public debt of country i measured in relation to (real) GDP. No seigniorage effects on governments' debt are assumed to be present. The interest rate on public debt (on bonds) is denoted by BI_{it}, which assumes an average bond maturity of 6 years, as estimated in Krause and Moyen (2013):

$$BI_{it} = \frac{1}{6}\sum_{\tau=t-5}^{t} I_{i\tau}. \tag{19}$$

Both national fiscal authorities are assumed to care about stabilizing inflation (π), output (y), debt (D), and fiscal deficits of their own countries (g) at each time t. This is a policy setting which seems plausible for the real EMU as well, with full employment (output at its potential level) and price level stability relating to country (or bloc) i's primary domestic goals, and government debt and deficit relating to its obligations according to the Maastricht Treaty of the European Union and its Stability and Growth Pact. The common central bank is interested in stabilizing inflation and output in the entire monetary union, also taking into account a goal of low and stable interest rates in the union. Hence, the individual objective functions of the national governments and of the common central bank are given by:

$$J_i = \frac{1}{2}\sum_{t=1}^{T}(\frac{1}{1+\theta})^{t-1}\{\alpha_{\pi i}(\pi_{it} - \tilde{\pi}_{it})^2 + \alpha_{yi}(y_{it} - \tilde{y}_{it})^2 + \alpha_{Di}(D_{it} - \tilde{D}_{it})^2 + \alpha_{gi}g_{it}^2\} \tag{20}$$

$$J_E = \frac{1}{2}\sum_{t=1}^{T}(\frac{1}{1+\theta})^{t-1}\{\alpha_{\pi E}(\pi_{Et} - \tilde{\pi}_{Et})^2 + \alpha_{yE}(y_{Et} - \tilde{y}_{Et})^2 + \alpha_E(R_{Et} - \tilde{R}_{Et})^2\} \tag{21}$$

where J_i $(i = 1, 2)$ denotes the objective function of the respective country or bloc and J_E denotes the objective function of the central bank.

The corresponding weights of the objective variables (their importance to the respective policy maker) are given in Table 1

Equations (11)–(18) constitute a dynamic game with three players, each of them having one control variable. The model contains 14 endogenous variables and four exogenous variables and is assumed to be played over a finite time horizon. The objective functions are quadratic in the paths of deviations of state and control variables from their respective desired values. The game is nonlinear-quadratic and hence cannot be solved analytically but only numerically. To this end, we have to specify the parameters of the model.

The parameters of the model are specified for a slightly asymmetric monetary union; see Table 2. Here an attempt has been made to calibrate the model parameters so as to fit for the Euro Area (EA). The data used for calibration include average economic indicators for the 16 EA countries from EUROSTAT up to the year 2007. Mainly based on the public finance situation, the EA is divided into two blocs of core (consisting of the following countries: Austria, Belgium, Estonia, Finland, France, Germany, Luxembourg, Malta, the Netherlands and Slovakia) and periphery (consisting of the following countries: Cyprus, Greece, Ireland, Italy, Portugal, Slovenia and Spain). The first bloc has a weight of 60 % in the entire economy of the monetary union (i.e. the parameter ω is equal to 0.6). The second bloc has a weight of 40 % in the economy of the union; it consists of countries with higher public debt and deficits and higher interest and inflation rates, on average. The weights correspond to the actual shares in EA real GDP. For the other parameters

Table 1 Weights of the objective variables

α_{yi}, α_{gi}	$\alpha_{\pi E}$	$\alpha_{yE}, \alpha_{\pi i}$	α_{D1}	α_{D2}	α_{RE}
1	2	0.5	0.01	0.0001	2.5

Table 2 Parameter values for an asymmetric monetary union, $i = 1, 2$

T	θ	ω	$\delta_i, \eta_i, \varepsilon_i$	$\beta_i, \gamma_i, \rho_i, \kappa_i, \lambda_i$	ξ_i	χ_i	μ_i, μ_E
30	3 %	0.6	0.5	0.25	0.1	0.0125	0.333

Table 3 Initial values of the two-country monetary union

$y_{i,0}$	$\pi_{i,0}$	$\pi^e_{i,0}$	$D_{1,0}$	$D_{2,0}$	$R_{E,0}$	$g_{1,0}$	$g_{2,0}$
0	2	2	60	80	3	0	0

Table 4 Target values for an asymmetric monetary union

$\tilde{D}_{1t}$	$\tilde{D}_{2t}$	$\tilde{\pi}_{it}$	$\tilde{\pi}_{Et}$	$\tilde{y}_{it}$	$\tilde{y}_{Et}$	$\tilde{g}_{it}$	$\tilde{R}_{Et}$
60	80↘60	2	2	0	0	0	3

of the model, we use values in accordance with econometric studies and plausibility considerations.

The initial values of the state variables of the dynamic game model are presented in Table 3. The ideal or target values assumed for the objective variables of the players are given in Table 4. Country 1 (the core bloc) has an initial debt level of 60 % of GDP and aims to hold this level over time. Country 2 (the periphery bloc) has an initial debt level of 80 % of GDP and aims to decrease its level to 60 % at the end of the planning horizon, which means that it is going to fulfil the Maastricht criterion for this economic indicator. The ideal rate of inflation is calibrated at 2 %, which corresponds to the Eurosystem's aim of keeping inflation below, but close to, 2 %. The initial values of the two blocs' government debts and budget deficits correspond to those at the beginning of the Great Recession, the recent financial and economic crisis. Otherwise, the initial situation is assumed to be close to equilibrium, with parameter values calibrated accordingly.

4 Optimal Macroeconomic Policies

4.1 Baseline Scenario

The MUMOD1 model can be used to simulate the effects of different shocks hitting the monetary union, which are reflected in the paths of the exogenous non-controlled variables, and of policy reactions towards these shocks. In this paper we introduce two sequences of different demand side shocks on the monetary union. The first sequence consists of a negative asymmetric demand side shock (zd_i), as given in Table 5, which aims at describing the macroeconomic dynamics of a monetary union in a situation similar to the economic crisis (2007–2010) and the ensuing sovereign debt crisis in Europe (since 2010). The values of the shock variables are calibrated so as to yield (pessimistic) estimates of the fall in GDP during these years. This scenario is used as a baseline for comparison with further experiments.

In the following figures, we show the time paths of the policy instruments and the endogenous variables for a simulation assuming constant values for the instruments ("simulation") and for the feedback Nash equilibrium solution ("Nash-FB") and the collusive Pareto-optimal solution ("Pareto") (Figs. 1 and 2).

In the baseline scenario, both countries suffer dramatically from the economic downturn modeled by the demand side shock in the first three periods. Output

Table 5 Negative asymmetric shock on demand side

t	1	2	3	4	5	6	7	8	9	...
zd_{1t}	−1	−6	−1	0	0	0	0	0	0	0
zd_{2t}	−1	−6	−1	−6	−8	−6	−4	−2	0	0

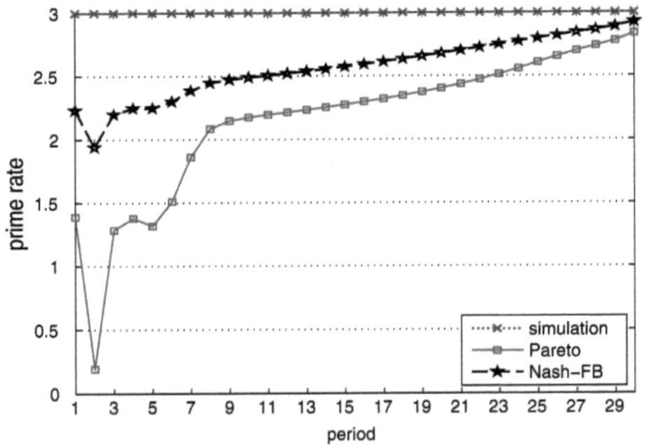

Fig. 1 Prime rate R_{Et} controlled by the central bank

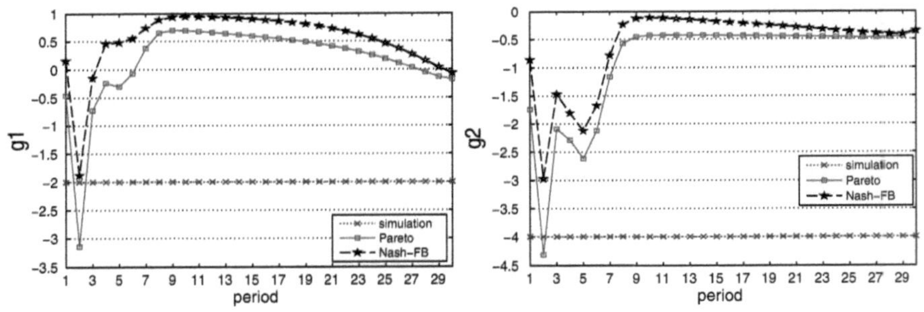

Fig. 2 Country i's fiscal surplus g_{it} (control variable) for $i = 1$ (core; *left*) and $i = 2$ (periphery; *right*)

drops by more than 6 %, which for several European countries is a fairly good approximation of what happened in reality. The periphery suffers additionally in periods 4–8 due to the second negative demand shock, which hits the second bloc only. Without policy intervention a combination of persistent budget deficits and an adverse economic environment leads to sky-rocketing public debts, which go up to 200 % of GDP for country 1 (core bloc) and 450 % for country 2 (periphery bloc). Without active policy intervention the second country at least would go bankrupt long before this level of public debt is achieved. Although such a solution may be regarded as unrealistic, it clearly shows the need for policy actions to stabilize the economies of the monetary union.

The calculated solutions of the baseline scenario imply that the optimal policies of both the governments and the common central bank are countercyclical during the immediate influence of the demand shock but not afterwards; instead, if governments want (or are obliged by the union's rules) to keep their public debt under control and avoid state bankruptcy, they have to implement prudent fiscal policies as soon as the crisis is over. The core bloc, which gives higher importance to the public debt target, follows this line and creates budget surpluses. In contrast, the periphery bloc runs a less prudent fiscal policy. As a result, the public debt of the periphery bloc goes up to 185 % of GDP in the case of the Pareto solution and up to 220 % of GDP in the case of the feedback Nash equilibrium solution.

The non-cooperative feedback Nash equilibrium and feedback Stackelberg equilibrium solutions give rather similar results. As this holds for all experiments considered, we refrain from showing the results of the feedback Stackelberg solution. In comparison to a Pareto optimal solution, the central bank acts less actively and the countries run more restrictive fiscal policies. As a result, output and inflation are slightly below the values achieved in the cooperative solution, and public debt is slightly above. In the Pareto solution the central bank cooperates and is willing to be more active in order to support the countries, which in turn may use fiscal policies to deal with the problem of the recessions by more expansionary activities (Figs. 3, 4, 5, 6, and 7).

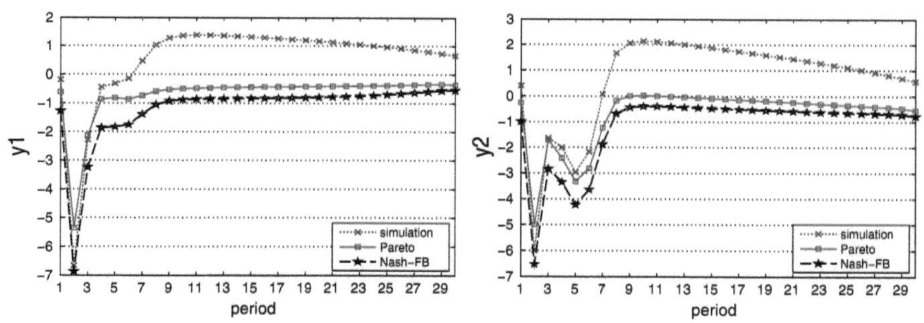

Fig. 3 Country i's output y_{it} for $i = 1$ (core; *left*) and $i = 2$ (periphery; *right*)

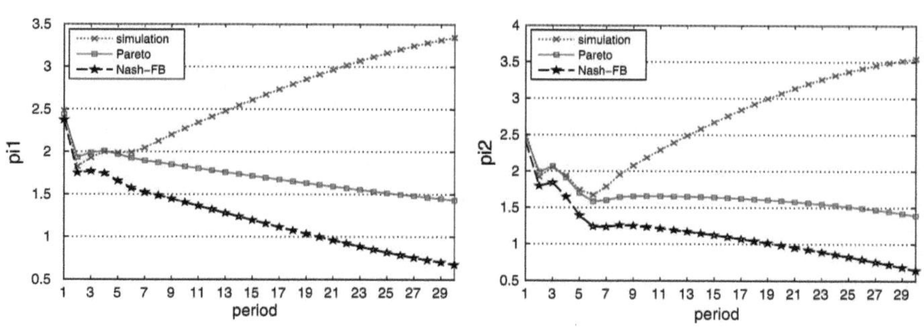

Fig. 4 Country i's inflation rate π_{it} for $i = 1$ (core; *left*) and $i = 2$ (periphery; *right*)

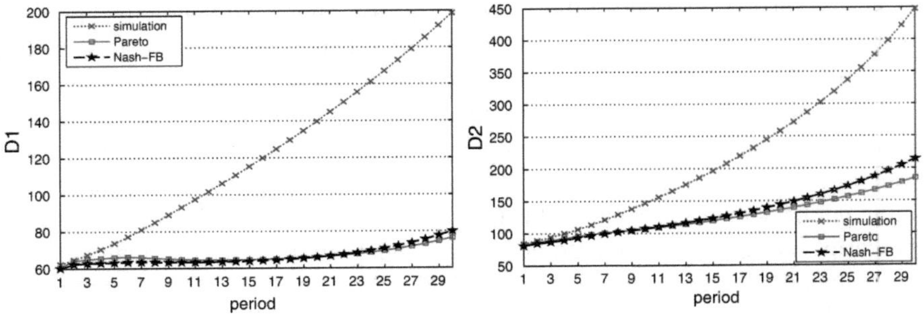

Fig. 5 Country i's debt level D_{it} for $i = 1$ (core; *left*) and $i = 2$ (periphery; *right*)

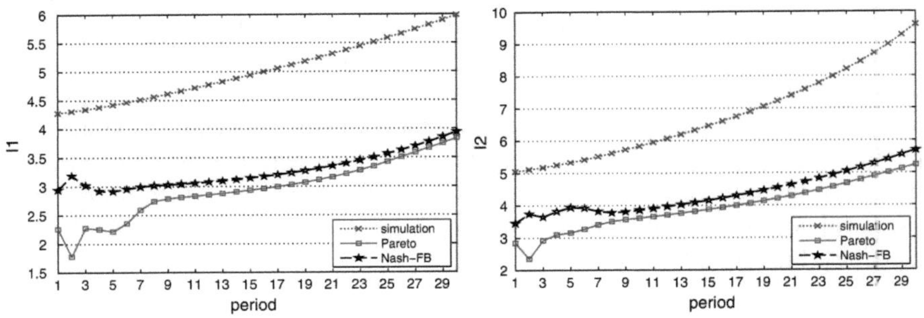

Fig. 6 Country i's nominal interest rate I_{it} for $i = 1$ (core; *left*) and $i = 2$ (periphery; *right*)

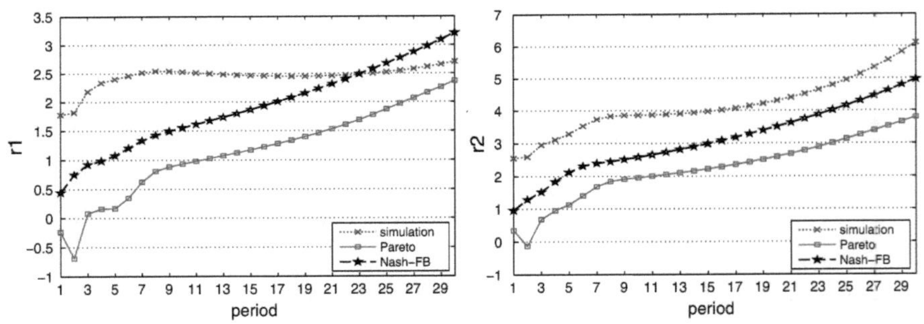

Fig. 7 Country i's real interest rate r_{it} for $i = 1$ (core; *left*) and $i = 2$ (periphery; *right*)

The second sequence of shocks serves to investigate the best macroeconomic policy strategies against different kinds of demand side shocks hitting the monetary union several periods after the initial demand side shock (starting in $t = 9$). We concentrate on three different shocks, a temporary, a reverse and a persistent one.

4.2 Impacts of a Temporary Demand Side Shock

In this section we consider a negative symmetric demand side shock (zd_i) which hits the system for two periods and then disappears as given in Table 6. It can be interpreted as a pessimistic current forecast for the (European) monetary union which, however, will lead to relatively quick recovery (in 2016 or so).

Both the time paths of the model variables in the uncontrolled simulation and in the game solutions are qualitatively similar to those of the first slump meant to model the Great Recession. Monetary policy lowers the prime rate, fiscal policies increase the budget deficit during the temporary second recession, and the consequences for output and debt are adverse. The cooperative solution is again superior to the noncooperative one, achieving higher output and lower public debt by a policy-mix of more expansionary fiscal and monetary policies. Especially the central bank is more active in the cooperative solution, leading to lower interest rates (nominal and especially real ones) and thus supporting investment and output growth (Figs. 8, 9, 10, 11, 12, 13, and 14).

Table 6 A temporary symmetric demand side shock

t	9	10	11	...
zd_{1t}	−5	−2	0	0
zs_{2t}	−5	−2	0	0

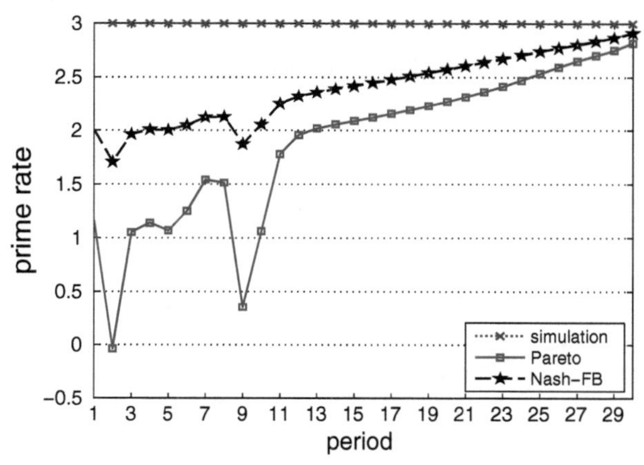

Fig. 8 Prime rate R_{Et} controlled by the central bank

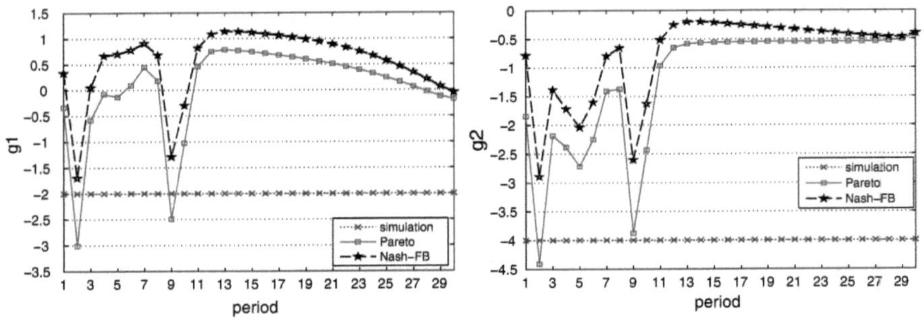

Fig. 9 Country i's fiscal surplus g_{it} (control variable) for $i = 1$ (core; *left*) and $i = 2$ (periphery; *right*)

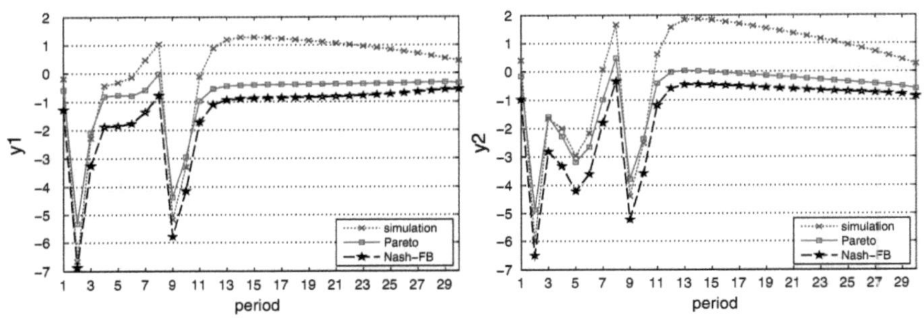

Fig. 10 Country i's output y_{it} for $i = 1$ (core; *left*) and $i = 2$ (periphery; *right*)

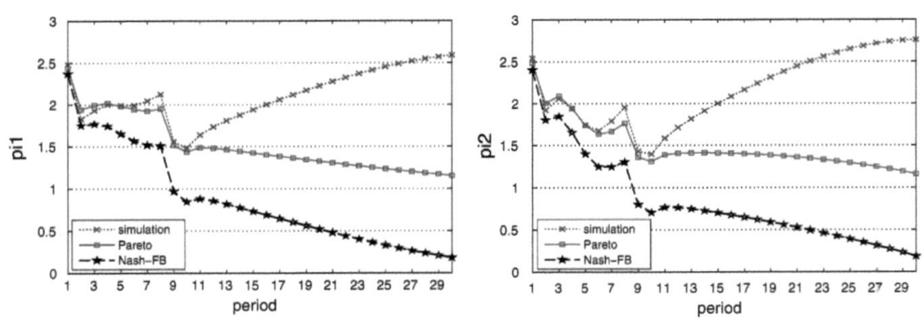

Fig. 11 Country i's inflation rate π_{it} for $i = 1$ (core; *left*) and $i = 2$ (periphery; *right*)

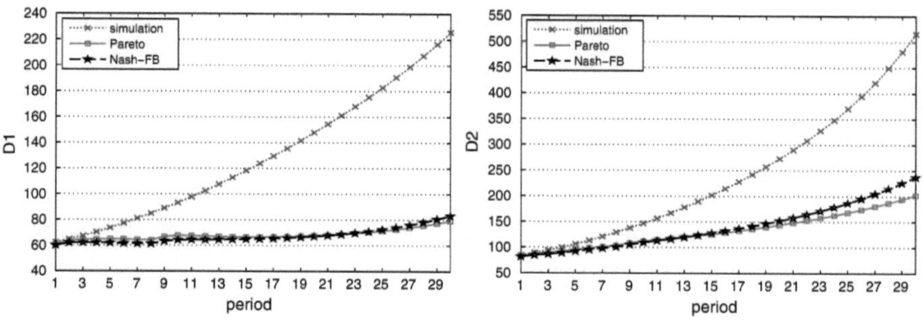

Fig. 12 Country i's debt level D_{it} for $i = 1$ (core; *left*) and $i = 2$ (periphery; *right*)

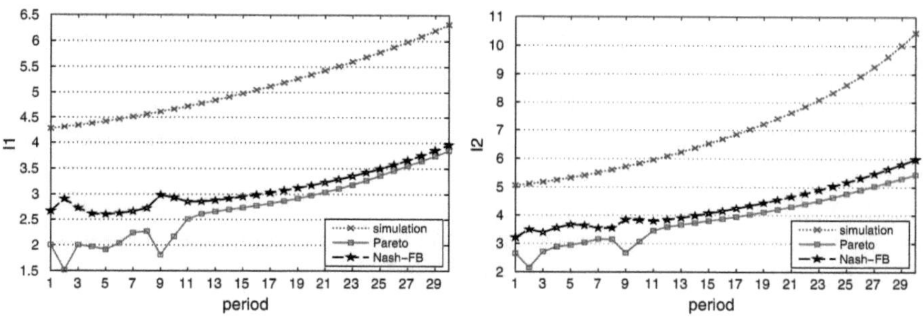

Fig. 13 Country i's nominal interest rate I_{it} for $i = 1$ (core; *left*) and $i = 2$ (periphery; *right*)

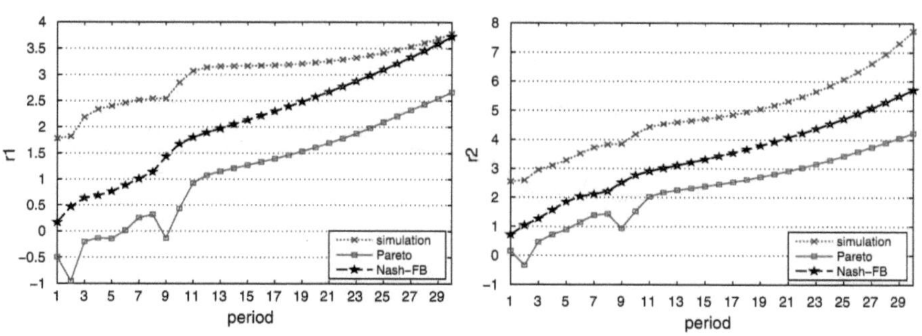

Fig. 14 Country i's real interest rate r_{it} for $i = 1$ (core; *left*) and $i = 2$ (periphery; *right*)

4.3 Impacts of a Reverse Demand Side Shock

Next, we consider a negative symmetric demand side shock (zd_i) which hits the system for two periods and is then followed by a positive shock of similar magnitude as given in Table 7.

The reversion of the negative demand side shock implies overshooting of (excess) output in both countries over an extended time period in the uncontrolled solution (the "simulation"). This means that the output (and, implicitly, the employment) objective becomes less pressing than in the previous scenarios. Interestingly, this has virtually no consequences for the design of monetary policy, which is nearly identical to the scenario with the temporary demand side shock without a recession. Fiscal policies differ, however, becoming more restrictive than before. Producing substantial (primary) surpluses during the boom period contributes to keeping the increase in public debt within reasonable limits, even in the periphery, which is less debt-averse than the core. Favorable effects on public debt also come from lower real interest rates, due both to lower nominal interest rates and higher inflation as compared to the scenario without reversion. It is also remarkable that the steady state path (with zero excess demand) is attained fairly quickly. The qualitative dynamic behavior of macroeconomic policies (countercyclical) and the differences between the noncooperative and the cooperative solution are comparable to those in the scenario without reversion (Figs. 15, 16, 17, and 18).

Table 7 A reverse symmetric demand side shock

t	9	10	11	12	13	14	15	...
zd_{1t}	−5	−2	2	2	2	1	0	0
zd_{2t}	−5	−2	2	2	2	1	0	0

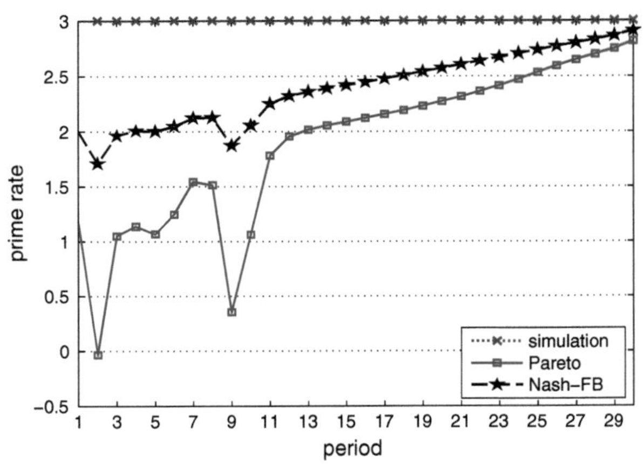

Fig. 15 Prime rate R_{Et} controlled by the central bank

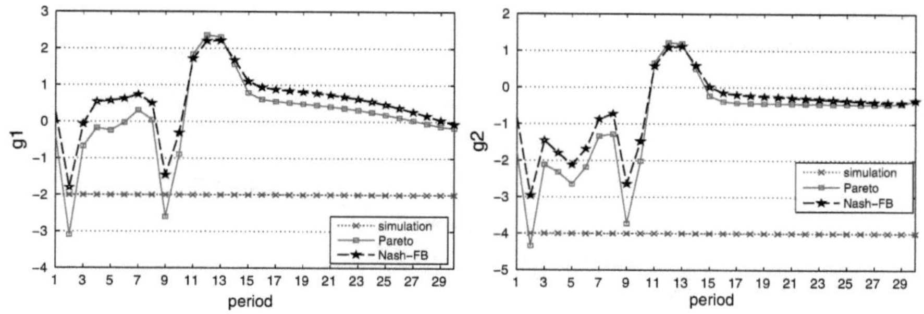

Fig. 16 Country i's fiscal surplus g_{it} (control variable) for $i = 1$ (core; *left*) and $i = 2$ (periphery; *right*)

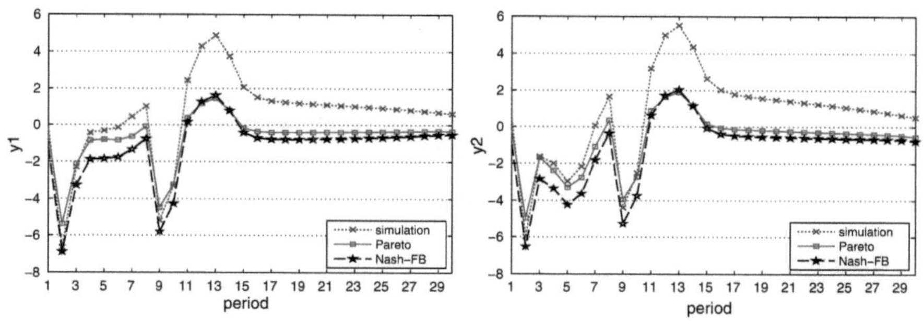

Fig. 17 Country i's output y_{it} for $i = 1$ (core; *left*) and $i = 2$ (periphery; *right*)

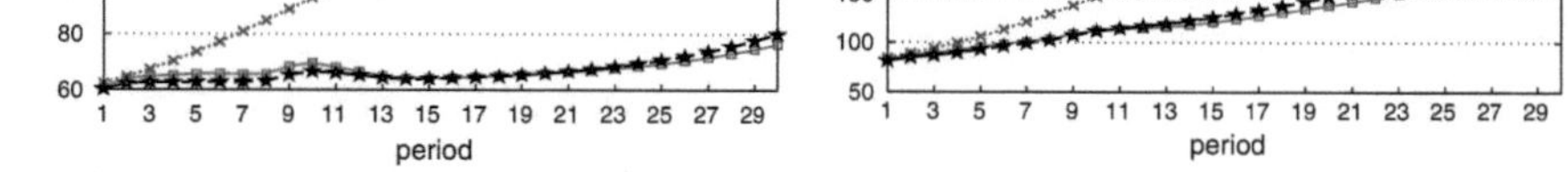

Fig. 18 Country i's debt level D_{it} for $i = 1$ (core; *left*) and $i = 2$ (periphery; *right*)

4.4 Impacts of a Persistent Demand Side Shock

Finally, we consider a negative symmetric demand side shock (zd_i) which hits the system for two periods and then diminishes but stays until the end of the planning horizon; see Table 8.

As Figs. 19, 20, 21, 22, and 23 show, in this case the trade-off between output and government debt becomes more challenging. Both the underutilization of output

Table 8 A persistent symmetric demand side shock

t	9	10	11	...	30
zd_{1t}	−5	−2	−0.5	−0.5	−0.5
zd_{2t}	−5	−2	−0.5	−0.5	−0.5

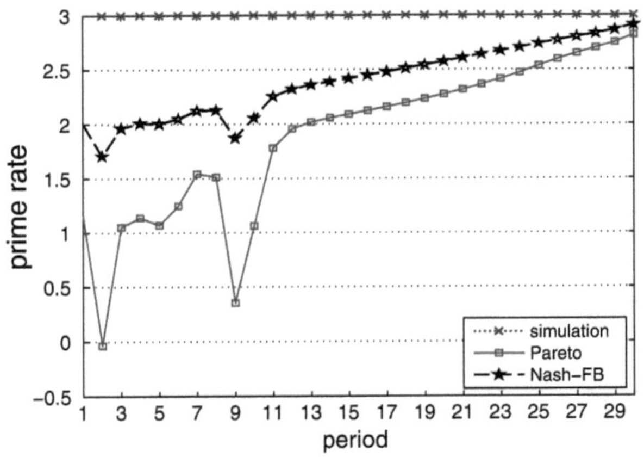

Fig. 19 Prime rate R_{Et} controlled by the central bank

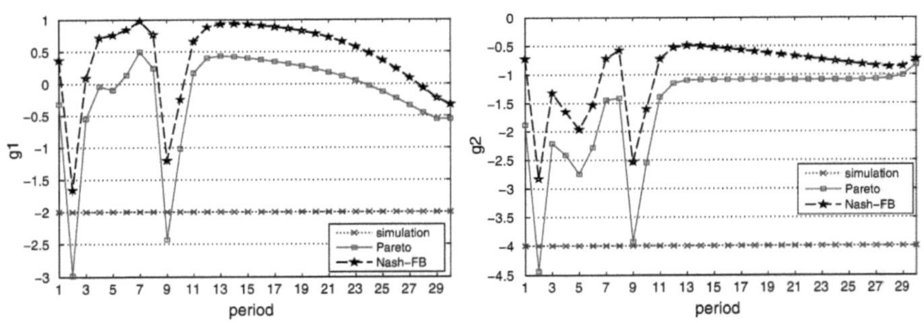

Fig. 20 Country i's fiscal surplus g_{it} (control variable) for $i = 1$ (core; *left*) and $i = 2$ (periphery; *right*)

and public debt are higher than in the other scenarios, especially that with reversion. Although the short run reactions of fiscal and monetary policies to this permanent shock are similar to those in the other scenarios—countercyclical expansion during the severe drop in aggregate demand—the long run trajectories show that increases in public debt and below natural output levels have to be traded off against each other. This results in a long run output level below zero (natural output) and a modest additional amount of government debt towards the end of the time horizon as compared to the other scenarios with the second negative demand side shock. This is also the only scenario which results in a (modest) deflation during the last periods; hence deflationary developments (often seen as a major danger of negative demand shocks) do not seem very likely in the context of the monetary union model under

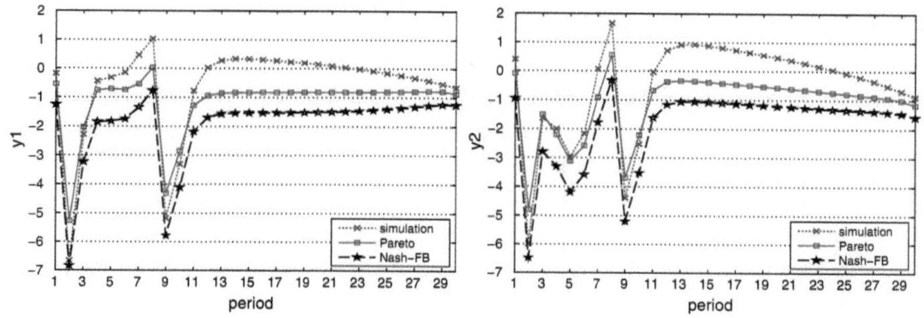

Fig. 21 Country i's output y_{it} for $i = 1$ (core; *left*) and $i = 2$ (periphery; *right*)

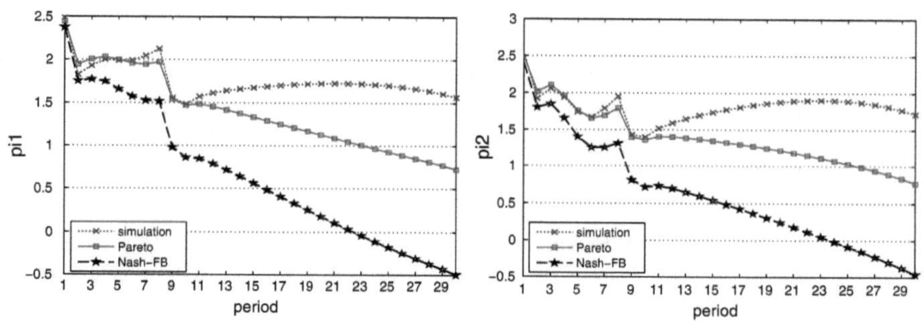

Fig. 22 Country i's inflation rate π_{it} for $i = 1$ (core; *left*) and $i = 2$ (periphery; *right*)

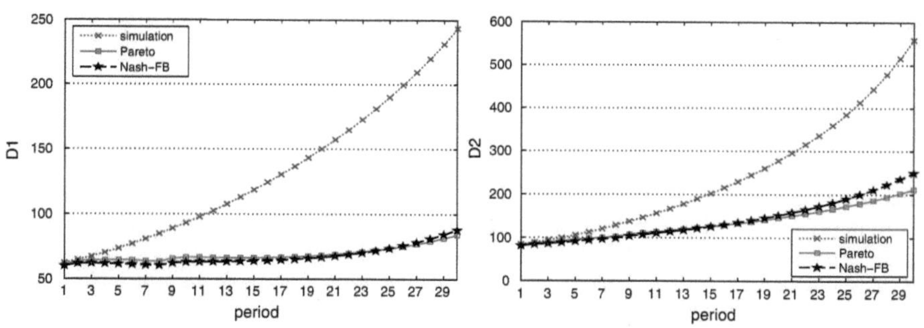

Fig. 23 Country i's debt level D_{it} for $i = 1$ (core; *left*) and $i = 2$ (periphery; *right*)

consideration. Finally, as in all other scenarios, the cooperative solution outperforms the noncooperative feedback Nash equilibrium solution. If we interpret the former as being based on some kind of "fiscal and monetary policy pact" or similar mechanism of compliance with rules in the monetary union, the advantage of such an agreement seems obvious—provided, of course, it is strong enough to become effective.

5 Concluding Remarks

In this paper we analyzed noncooperative equilibrium and cooperative strategies of fiscal and monetary policy makers in a macroeconomic model of a monetary union consisting of two blocs (treated as countries), the core and the periphery. The model was calibrated so as to mirror some essential aspects of the Euro Area from the Great Recession to possible future developments. In particular, various negative demand side shocks and their effects on key macroeconomic variables and on optimal policy strategies were examined; a temporary, a reverse, and a persistent drop in exogenous aggregate demand affecting the entire monetary union. These possible future shocks were superimposed on shocks serving as models for the Great Recession and the sovereign debt crisis in Europe. The results show that such shocks provide a challenge for the policy makers of the monetary union, with a tradeoff especially between the requirement of keeping output (and, implicitly, employment) close to its natural levels and the avoidance of excessive government debt threatening the solvency of the union, in particular its periphery bloc. The policies called for are countercyclical and mildly expansionary; only under a permanent shock will the model economy not be able to return to its steady state path. In all scenarios, the cooperative solution dominates the noncooperative equilibrium solution, which may be interpreted as an argument in favor of an enforceable pact between governments and the common central bank.

References

Basar T, Olsder GJ (1999) Dynamic noncooperative game theory, 2nd edn. SIAM, Philadelphia

Blueschke D, Neck R (2011) "Core" and "periphery" in a monetary union: a macroeconomic policy game. Int Adv Econ Res 17:334–346

Blueschke D, Neck R, Behrens DA (2013) OPTGAME3: a dynamic game solver and an economic example. In: Krivan V, Zaccour G (eds) Advances in dynamic games - theory, applications, and numerical methods, pp 29–51. Birkhäuser Verlag, Basel

Dixit A, Lambertini L (2001) Monetary-fiscal policy interactions and commitment versus discretion in a monetary union. Eur Econ Rev 45:977–987

Farmer K (2006) Reducing public debt under dynamic efficiency: transitional dynamics in Diamond's OLG model. Atl Econ J 34(2):195–208

Farmer K (2011) Public-debt sustainability, real exchange rate, and country-specific saving rates. Int Adv Econ Res 17(1):45–65

Farmer K, Zotti J (2010a) Sustainable government debt in a two-country overlapping generations' model. Int Adv Econ Res 16(1):124–125

Farmer K, Zotti J (2010b) Sustainable government debt in a two-good, two-country overlapping generations' model. Int Rev Econ 57(3):289–316

Farmer K, Schelnast M (2013) Public debt reduction in advanced countries and its impact on emerging countries. Int Adv Econ Res 19(2):167–188

Haber G, Neck R, McKibbin WJ (2002) Global implications of monetary and fiscal policy rules in the EMU. Open Econ Rev 13:363–379

Krause MU, Moyen S (2013) Public debt and changing inflation targets. Bundesbank discussion paper 6/2013

Neck R, Behrens DA (2009) A macroeconomic policy game for a monetary union with adaptive expectations. Atl Econ J 37:335–349

Neck R, Blueschke D (2014) "Haircuts" for the EMU periphery: virtue or vice? Empirica 41:153–175

van Aarle B, Di Bartolomeo G, Engwerda J, Plasmans J (2002) Monetary and fiscal policy design in the EMU: an overview. Open Econ Rev 13:321–340.

Determinants of Maximum Sustainable Government Debt

Anna Boisits and Matthias Schelnast

1 Introduction

Karl Farmer is one of a few people at the University of Graz who introduced us to the impressive sphere of overlapping generations (OLG) models. These kinds of models are useful for analyzing economic problems in cases in which international interrelations as well as different effects on coexisting generations play a role. Much of his research has been dedicated to the fields of climate policy and government debt. This article is devoted to the latter.

It is well known that Ricardian equivalence holds in an infinitely lived agent (ILA) model with rational agents and perfect information. Hence, we have to look for another kind of model, one which is capable to account for the non-neutrality of public debt. OLG models incorporating government debt were pioneered by Diamond (1965). Diamond demonstrated that in a simple one-country one-good economy, public debt may have a positive or a negative steady-state welfare effect depending on the dynamic inefficiency or efficiency of the economy. Based on this research, Rankin and Roffia (2003) have shown that limits to government debt in such a model exist where the maximum steady-state debt ratio is at a point where the capital stock is neither zero nor maximized. For the case of internationally integrated economies, Farmer and Zotti (2010) proved that such limits also occur in a two-country two-good framework. However, maximum debt ratios seem to differ quite radically between countries. Whereas Japan has maintained a debt–GDP ratio of

A. Boisits (✉)
Center for Accounting Research, University of Graz, Universitätsstraße 15, A-8010 Graz, Austria
e-mail: anna.boisits@uni-graz.at

M. Schelnast
Department of Economics, University of Graz, Universitätsstraße 15, A-8010 Graz, Austria
e-mail: schelnast@gmx.at

B. Bednar-Friedl, J. Kleinert (eds.), *Dynamic Approaches to Global Economic Challenges*, DOI 10.1007/978-3-319-23324-6_5

more than 200 % for several years, significantly lower debt ratios in some European countries have caused substantial difficulties for the respective countries.

Farmer and Zotti (2010) evidenced a negative relationship between domestic debt limits and foreign government debt. Schelnast (2013) extended the basic model by introducing productive government expenditures. He investigated the effect of some economic variables, such as savings rates or productive government expenditures,[1] on maximum debt ratios. However, this was done in a two-country two-good model, in which home and foreign countries exhibit similar technologies. We extend the research of Schelnast (2013) by introducing country-specific technologies. Therefore, the purpose of our model is to capture international interdependencies between country groups at different stages of development, such as China/India on the one hand and Europe/the USA on the other. First, we analyze the effect of economic variables on the home country's debt limits in a setting with two country groups with different technologies; these variables include the foreign country's debt ratio, public productive expenditure, and savings rates. Farmer and Zotti (2010) and Schelnast (2013) discussed these relationships in a setting with country groups of similar development. Beyond confirming their results for a setting with different technologies, we study the effect of higher unproductive government spending and an increase in the output elasticity of capital on the home country's maximum debt limit.

The remainder of this article is organized as follows: The next section provides a brief overview of the OLG model setting used in the paper, and defines the intertemporal and steady-state equilibrium conditions. The following section shows the dependencies of the home country's debt limits and various economic variables. Finally, the last section concludes and sums up the main findings of our analysis.

2 The Model

We consider an OLG model in which two generations live at the same time. As in Farmer and Zotti (2010), we consider two integrated countries or country groups, i.e., Europe/the USA and China/India, which we call Home (Europe/the USA) and Foreign (China/India). Subsequently, Foreign variables are labeled with a superscript asterisk. The two country groups differ in terms of total factor productivity as well as capital and labor productivity.

It is assumed that households live over two periods. 'Young' households consist of two (young) adults and their children and 'old' households comprise two (old) adults. The children of the previous period make up the young adults in the present period. Supposing that the typical period length is 25–30 years, this implies that a

[1]Some rationales for considering productive government expenditures in the analysis of debt and debt limits are given in Schelnast (2013, p. 5ff).

single individual lives for 75–90 years, while households have a lifetime of 50–60 years.

The adults of the younger generation constitute the workforce of a country. At time t, there are L_t young households. To account for the empirical fact of population growth in most parts of the world, this number is assumed to grow at a constant growth factor A^L. Thus, the number of young households at $t+1$ is given by $L_{t+1} = A^L L_t$. This generation provides labor to firms inelastically of the wage rate. When old, the members of the households retire and live on their private savings.

In our economy, each country (or country group) produces a country-specific bundle of goods. For simplicity, we consider a pure exchange economy, in which Home's (Foreign's) variables are valuated in Home's (Foreign's) good. Hence, the real exchange rate, e_t, denotes the amount of Home's good received in exchange for one unit of the Foreign good.

In addition to households, which make consumption and savings decisions, and firms, which produce consumption and investment goods, the government is the third main player in our model. The governments of Home and Foreign collect (income) taxes, issue debt to finance productive and unproductive public expenditures, and pay back mature debt.

In the following subsections, we explain the decision problems of each of these groups.

2.1 Households

Young and old households can decide on their consumption of Home's good, x_t, and Foreign's good, y_t. We assume that the utility of the representative young household born in period t in Home is given by the following log-linear utility function:

$$\begin{aligned} U_t = {} & \zeta \ln x_t^1 + (1-\zeta)\ln y_t^1 + \gamma \ln \frac{G_t^U}{L_t + L_{t-1}} \\ & + \beta \left[\zeta \ln x_{t+1}^2 + (1-\zeta)\ln y_{t+1}^2 + \gamma \ln \frac{G_{t+1}^U}{L_{t+1} + L_t} \right]. \end{aligned} \tag{1}$$

Households have to decide between present and future consumption. $0 < \beta < 1$ and $0 < \beta^* < 1$ reflect the time preference of consumption in the first period versus consumption in the second period in Home and Foreign. x_t^1 (y_t^1) denotes the household consumption level of the good produced in Home (Foreign) when young. The superscript 2 denotes the respective consumption in the retirement period. Households' bias towards Home products is taken into account through the utility weighting $0 < \zeta < 1$. In addition, we assume that consumers gain utility from the consumption of a governmental good. Consumers' valuation of this governmental good is reflected by the weight γ.

Public expenditures, which only increase household utility but do not enhance productivity, are given by G_t^U, where the U in the superscript illustrates that this expenditure is unproductive. The governmental good is assumed to be rival, and thus only the governmental good per person enters the utility function. At time t, the number of households in Home is given by the young and old households living in this period: $L_t + L_{t-1}$. Thus, unproductive expenditure per household at time t is given by $\left(G_t^U / (L_t + L_{t-1})\right)$.

Likewise, Foreign household utility is represented by:

$$\begin{aligned} U_t^* &= \zeta \ln x_t^{*,1} + (1-\zeta) \ln y_t^{*,1} + \gamma \ln \frac{G_t^{*,U}}{L_t^* + L_{t-1}^*} \\ &\quad + \beta^* \left[\zeta \ln x_{t+1}^{*,2} + (1-\zeta) \ln y_{t+1}^{*,2} + \gamma \ln \frac{G_{t+1}^{*,U}}{L_t^* + L_{t+1}^*} \right], \end{aligned} \tag{2}$$

where we assume that households in Home and Foreign differ in their time preference, $\beta^* \neq \beta$.

Young households receive a gross salary w_t for their work and have to pay taxes. The percentage labor income tax is given by τ_t. The disposable income, $(1-\tau_t)\, w_t$, can then either be spent on the consumption of the goods produced in Home or Foreign, or can be saved. Savings per young household born at time t are given by s_t. Savings can be in the form of investments in real capital goods (i.e., machinery, real estate, etc.) or bonds issued by Home's or Foreign's government. It is assumed that real capital is immobile and bonds are perfectly mobile, implying that households can purchase only domestic real investment goods but domestic and foreign bonds. Without loss of generality, we assume that real capital depreciates fully in the course of one period. Thus, current investments in real capital coincide with the next period's available real capital stock, K_{t+1}. Therefore, the budget constraints of young households living in Home are given by:

$$x_t^1 + e_t y_t^1 + s_t = (1-\tau_t)\, w_t, \text{ where } s_t \equiv K_{t+1}/L_t + B_{t+1}^H/L_t + e_t B_{t+1}^{*,H}/L_t. \tag{3}$$

Here B_{t+1}^H ($B_{t+1}^{*,H}$) gives the total amount of bonds issued by Home's (Foreign's) government held by Home's households.

In its retirement period, a household lives on its savings plus interest payments from real capital goods and domestic and foreign bonds. The return on Home's real capital is denoted by $(q_t - 1)$.

The budget constraints in the retirement period are given by:

$$x_{t+1}^2 + e_{t+1} y_{t+1}^2 = q_{t+1} \left(K_{t+1}/L_t + B_{t+1}^H/L_t\right) + q_{t+1}^* e_{t+1} \left(B_{t+1}^{*,H}/L_t\right). \tag{4}$$

In the absence of uncertainty about yields, real capital and Home's bonds have to generate the same rate of return so that households invest in real capital as well as in bonds. Furthermore, Home's households demand both Foreign's and Home's government bonds only if they get the same reward for these investment opportunities. Thus, households are only indifferent between domestic and foreign bonds if the following equation holds:

$$q_{t+1} = (e_{t+1}/e_t)\, q^*_{t+1}. \tag{5}$$

Equation (5) is an international interest parity condition.

The goal of each household is to choose the consumption quantities in the working and the retirement period that maximize its utility (1) given the budget constraints (3) and (4). Solving this constraint maximization problem leads to the optimal consumption and savings values:

$$\begin{aligned} x_t^1 &= \tfrac{\zeta}{1+\beta}\left[(1-\tau_t)\, w_t\right],\ y_t^1 = \tfrac{1-\zeta}{(1+\beta)e_t}\left[(1-\tau_t)\, w_t\right],\ x_{t+1}^2 = \tfrac{\beta\zeta}{1+\beta} q_{t+1}\left[(1-\tau_t)\, w_t\right], \\ y_{t+1}^2 &= \frac{\beta(1-\zeta)}{(1+\beta)\, e_{t+1}} q_{t+1}\left[(1-\tau_t)\, w_t\right] \text{ and } s_t = \sigma\left[(1-\tau_t)\, w_t\right], \text{ where } \sigma \equiv \frac{\beta}{1+\beta}. \end{aligned} \tag{6}$$

The optimal consumption and savings decisions given in (6) imply that demand for all goods increase with disposable income. The more income a household has, the more of everything it buys. Moreover, the greater the patience of a household, the higher its savings, and hence the higher its consumption will be in the retirement period. Clearly, the savings rate, σ, is exogenous and only determined by the time preference of households, β. Finally, the more a household prefers Home's good compared to Foreign's good, the higher (lower) the demand for Home's (Foreign's) good.

The same approach can be applied to calculate the optimal consumption and savings rates for Foreign's households:

$$\begin{aligned} x_t^{*,1} &= \frac{\zeta}{1+\beta^*} e_t\left[\left(1-\tau_t^*\right) w_t^*\right],\ y_t^{*,1} = \frac{1-\zeta}{1+\beta^*}\left[\left(1-\tau_t^*\right) w_t^*\right], \\ x_{t+1}^{*,2} &= \frac{\beta^*\zeta}{1+\beta^*} e_{t+1} q^*_{t+1}\left[\left(1-\tau_t^*\right) w_t^*\right], \\ y_{t+1}^{*,2} &= \frac{\beta^*(1-\zeta)}{1+\beta^*} q^*_{t+1}\left[\left(1-\tau_t^*\right) w_t^*\right] \text{ and } s_t^* = \sigma^*\left[\left(1-\tau_t^*\right) w_t^*\right], \text{ where} \\ \sigma^* &\equiv \frac{\beta^*}{1+\beta^*}. \end{aligned} \tag{7}$$

2.2 Government

Government expenses in period t consist of three parts; productive government expenditures, G_t^G, unproductive government expenditures, G_t^U, and the repayment of expired government bonds plus interest payments, $q_t B_t$. An expenditure is called 'productive' if it increases the total factor productivity of a country. Examples are expenditure on infrastructure or public enterprises. An expenditure is 'unproductive' if an increase in the provision of this good does not have an impact on the productivity of the respective country. However, these expenditures may increase household utility. Examples might be expenditures for defense or law and order, culture, and environmental quality.

To fund its expenditures, the government has to obtain revenues. On the one hand, the government levies income taxes, $\tau_t w_t L_t$; on the other hand, it issues new bonds, B_{t+1}. Thus, the budget constraints of Home's (Foreign's) government are given by:

$$B_{t+1} + \tau_t w_t L_t = q_t B_t + G_t^G + G_t^U \left(B_{t+1}^* + \tau_t^* w_t^* L_t^* = q_t^* B_t^* + G_t^{*,G} + G_t^{*,U}\right). \tag{8}$$

We assume, as in Farmer (2006), that the goal of the government is to hold the expenditure share as a percentage share of Home's output constant. Therefore, government expenditures are given by:

$$G_t^G = gX_t, \; G_t^{*,G} = g^* Y_t, \; G_t^U = uX_t \text{ and } G_t^{*,U} = u^* Y_t \tag{9}$$

where X_t (Y_t) is Home's (Foreign's) total output; g (g^*) gives the percentage of output spent on productive governmental expenditures; u (u^*) give the percentage of output spent on unproductive governmental expenditures in Home (Foreign).

Furthermore, it is assumed that both governments pursue a constant stock fiscal policy (for further detail based on a similar assumption, see De la Croix and Michel 2002). This implies that the debt to output ratios are kept constant. Hence, $b = B_t/X_t = B_{t+1}/X_{t+1}$ and $b^* = B_t^*/Y_t = B_{t+1}^*/Y_{t+1}$. This assumption can be justified, for example, by the fact that the Maastricht Treaty dictates a certain debt–GDP ratio to European countries. Countries have to achieve and subsequently maintain the required debt ratio. We demonstrate in Sect. 3 that the maximum debt ratio that can be maintained by a specific country depends on various economic variables.

The constant stock fiscal policy implies that the tax rates in Home and Foreign become endogenous. Rearranging the government's budget constraints and taking account the constant stock fiscal policy and constant expenditure shares leads to:

$$\tau_t = \frac{X_t}{w_t L_t}\left[b\left(q_t - \frac{X_{t+1}}{X_t}\right) + (g+u)\right] \text{ and}$$
$$\tau_t^* = \frac{Y_t}{w_t^* L_t^*}\left[b^*\left(q_t^* - \frac{Y_{t+1}}{Y_t}\right) + (g^* + u^*)\right]. \tag{10}$$

Thus, a constant stock fiscal policy implies that the tax rate must be adjusted over time to hold the debt ratio constant.

2.3 Firms

Firms use three inputs to produce their output: labor, capital, and productive government expenditures. We assume that the production of goods in the two economies can be described by a Cobb–Douglas production function with constant returns to scale. Furthermore, we assume that government expenditures are rival. Therefore, total production depends on government expenditures per young household, $G_t^S = G_t^G/L_t$ and $G_t^{*,S} = G_t^{*,G}/L_t^*$. Likewise, total factor productivity in Home and Foreign, M and $M*$, may differ between countries.

$$X_t = M\left(G_t^S\right)^{\alpha^G}(L_t)^{1-\alpha}(K_t)^{\alpha} \text{ and } Y_t = M^*\left(G_t^{*S}\right)^{\alpha^{*G}}\left(L_t^*\right)^{1-\alpha*}\left(K_t^*\right)^{\alpha*}, \tag{11}$$

where $\alpha^G + \alpha < 1$ and $\alpha^{G*} + \alpha^* < 1$.

By inserting (9) into (11), and defining output per young household by $x_t = X_t/L_t$, and capital per young household (capital intensity) by $k_t = K_t/L_t$ and $\tilde{g} \equiv g^{\alpha^G/(1-\alpha^G)}M^{1/(1-\alpha^G)}$, output per young household is given by:

$$x_t = \tilde{g}k_t^{\frac{\alpha}{1-\alpha^G}} \text{ and } y_t = \tilde{g}^*\left(k_t^*\right)^{\frac{\alpha*}{1-\alpha^{*G}}}. \tag{12}$$

The availability of productive government expenditures is exogenous for private firms. Capital and labor, on the other hand, are at the discretion of the firm. Firms choose these variables to maximize profits. Hence, real wage and rental rates are given by the partial derivative of the production functions given in (11):

$$w_t = (1-\alpha)\,\tilde{g}k_t^{\frac{\alpha}{1-\alpha^G}} \text{ and } w_t^* = \left(1-\alpha^*\right)\tilde{g}^*\left(k_t^*\right)^{\frac{\alpha^*}{1-\alpha^{*G}}}, \tag{13}$$

$$q_t = \alpha\tilde{g}k_t^{\frac{\alpha+\alpha^G-1}{1-\alpha^G}} \text{ and } q_t^* = \alpha^*\tilde{g}^*\left(k_t^*\right)^{\frac{\alpha^*+\alpha^{*G}-1}{1-\alpha^{*G}}}. \tag{14}$$

2.4 Intertemporal Equilibrium

In an intertemporal equilibrium the interest parity condition holds, and both product and financial markets are in equilibrium.

First, we consider the product markets of Home and Foreign goods. Home's production in period t is used either for the consumption of the young and old Home's or Foreign's households, for Home's private investments in real capital, or for Home's government expenditures. The same is true for Foreign's good. Thus, product markets are in equilibrium if the following equations hold:

$$X_t = L_t\left(x_t^1 + x_t^2/A^L\right) + L_t^*\left(x_t^{*,1} + x_t^{*,2}/A^L\right) + K_{t+1} + G_t^G + G_t^U, \tag{15}$$

$$Y_t = L_t\left(y_t^1 + y_t^2/A^L\right) + L_t^*\left(y_t^{*,1} + y_t^{*,2}/A^L\right) + K_{t+1}^* + G_t^{*,G} + G_t^{*,U}. \tag{16}$$

Second, it should be noted that households are only indifferent between buying Home's and Foreign's bonds if the international interest parity condition holds, as given in (5). This condition, together with the equations in (14), leads to:

$$e_{t+1} = \frac{\alpha\tilde{g}}{\alpha^*\tilde{g}^*} \frac{k_{t+1}^{\frac{\alpha+\alpha^G-1}{1-\alpha^G}}}{\left(k_{t+1}^*\right)^{\frac{\alpha^*+\alpha^{*G}-1}{1-\alpha^{*G}}}} e_t. \tag{17}$$

Third, the financial market has to be in equilibrium, which is the case if the supply of investment products equals demand. The total amount of savings of Home's households is invested in real capital goods and Home's and Foreign's bonds: $L_t s_t \equiv K_{t+1} + B_{t+1}^H + e_t B_{t+1}^{*,H}$. The same is true for Foreign's savings. Furthermore, total bonds issued by Home and Foreign have to be equal to total demand for these bonds:

$$B_t = B_t^H + B_t^F \text{and } B_t^* = B_t^{*,H} + B_t^{*,F}. \tag{18}$$

Adding up the total savings of Home's and Foreign's households measured in Home goods, and taking into account that bonds markets are in equilibrium, leads to the following equation:

$$s_t L_t + e_t s_t^* L_t^* = K_{t+1} + B_{t+1} + e_t \left(K_{t+1}^* + B_{t+1}^*\right). \tag{19}$$

Equations (15)–(19) define the intertemporal equilibrium. This system of equations can be simplified as follows: The product market clearing Eqs. (15) and (16) can be merged into one equation. First, in equilibrium, Home's and Foreign's households maximize their utility, and thus we can use the optimal consumption quantities given in (6) and (7) to simplify Eqs. (15) and (16). Next, we multiply Eq. (16) by $[\zeta/(1-\zeta)]\, e_t$ and add the left hand side of the transformed Eq. (16) to the right-hand side of Eq. (15), and vice versa. Dividing this new equation by L_t and defining $\rho \equiv L_t^*/L_t$, we obtain a single equation characterizing product market clearing:

$$\begin{aligned} &(1-(g+u))\,\tilde{g}k_t^{\frac{\alpha}{1-\alpha^G}} - \frac{\zeta}{(1-\zeta)}\rho e_t \left(1-(g^*+u^*)\right)\tilde{g}^*\left(k_t^*\right)^{\frac{\alpha^*}{1-\alpha^{*G}}} \\ &= A^L k_{t+1} - \frac{\zeta}{(1-\zeta)}\rho e_t A^L k_{t+1}^* \end{aligned} \tag{20}$$

Furthermore, the dynamic equations for the labor income tax rates in (10) can be reformulated:

$$\begin{aligned} \tau_t &= \tfrac{1}{1-\alpha}\left[b\left(q_t - A^L\left(\tfrac{k_{t+1}}{k_t}\right)^{\frac{\alpha}{1-\alpha^G}}\right) + (g+u)\right] \text{ and} \\ \tau_t^* &= \tfrac{1}{1-\alpha^*}\left[b^*\left(q_t^* - A^L\left(\tfrac{k_{t+1}^*}{k_t^*}\right)^{\frac{\alpha^*}{1-\alpha^{*G}}}\right) + (g^*+u^*)\right]. \end{aligned} \tag{21}$$

The financial market clearing condition, Eq. (19), can be rearranged using the optimal savings rates given in (6) and (7) and the tax rates given in (21).

$$\sigma\left(\left(1-\alpha-(g+u)\right)-bq_t\right)\tilde{g}k_t^{\frac{\alpha}{1-\alpha^G}}+\rho e_t\sigma^*\left(\left(1-\alpha^*-(g^*+u^*)\right)-b^*q_t^*\right)$$
$$\tilde{g}^*\left(k_t^*\right)^{\frac{\alpha^*}{1-\alpha^{*G}}}=$$
$$=A^Lk_{t+1}+b\left(1-\sigma\right)A^L\tilde{g}k_{t+1}^{\frac{\alpha}{1-\alpha^G}}+\rho e_t\left(A^Lk_{t+1}^*+b^*\left(1-\sigma^*\right)A^L\tilde{g}^*\left(k_{t+1}^*\right)^{\frac{\alpha^*}{1-\alpha^{*G}}}\right) \tag{22}$$

To sum up, the intertemporal equilibrium is fully described by the financial market clearing condition stated in Eq. (22), by the product market clearing condition in Eq. (20), and by the interest parity condition in Eq. (17).

2.5 *Steady State Equilibrium*

The previous section has shown that an intertemporal equilibrium is described by equilibrium in the financial market (Eq. (22)) and in the product market (20), and by the interest parity condition (17). The purpose of this section is to define the steady state equilibrium of our model. In such an equilibrium, the endogenous variables stay constant over time: $k_{t+1} = k_t = k$, $k_{t+1}^* = k_t^* = k^*$ and $e_{t+1} = e_t = e$. We will show that the endogenous dynamic variables, capital per young households and the real interest rate, converge towards finite levels if steady state equilibrium exists. The reason is that we assume decreasing returns to scale of private capital and public spending. A steady state equilibrium can be derived by solving the three equations, (17), (20), and (22), with respect to Home's and Foreign's capital stock and the real interest rate under the assumption that the variables are unchanging in time.

In steady state equilibrium, the dynamic equation for the real exchange rate (17) is simplified to:

$$k^*=\left(\frac{\alpha\tilde{g}}{\alpha^*\tilde{g}^*}\right)^{\frac{1-\alpha^{*G}}{\alpha^*+\alpha^{*G}-1}}k^{\frac{\left(\alpha+\alpha^G-1\right)\left(1-\alpha^{*G}\right)}{\left(1-\alpha^G\right)\left(\alpha^*+\alpha^{*G}-1\right)}}=\mu k^\eta, \tag{23}$$

with $\eta \equiv \left[\left(\alpha+\alpha^G-1\right)\left(1-\alpha^{*G}\right)\right]/\left[\left(1-\alpha^G\right)\left(\alpha^*+\alpha^{*G}-1\right)\right]$ and $\mu \equiv \left[\left(\alpha\tilde{g}\right)/\left(\alpha^*\tilde{g}^*\right)\right]^{\left(1-\alpha^{*G}\right)/\left(\alpha^*+\alpha^{*G}-1\right)}$. Equation (23) indicates that Foreign's steady state capital stock per labor is increasing in Home's steady state capital stock per young households. Higher capital intensity for Home implies a lower rental rate in Home. Because the interest parity condition holds, the rental rate in Foreign must also be low. Foreign's rental rate decreases if Foreign's capital intensity increases. Furthermore, the higher Home's total factor productivity, M, and the higher Home's government expenditure share, g, the lower Foreign's capital per labor compared to

Home's capital intensity. The opposite is true for Foreign's factor productivity, $M*$, and Foreign's government expenditure share, $g*$.

By using the relationship between Home's and Foreign's capital intensity (Eq. (23)), the equation for product market clearing, (20), can be simplified to:

$$e = \frac{1-\zeta}{\zeta}\frac{1}{\rho\mu}\frac{H}{H^*}, \tag{24}$$

where $H \equiv (1-(g+u))\,\tilde{g}k^{\alpha/(1-\alpha^G)} - A^L k$ and $\mu H^* \equiv \mu\Big[(1-(g^*+u^*))\,(\alpha/\alpha^*)\,\tilde{g}k^{\eta\alpha^*/(1-\alpha^{*G})} - A^L k^\eta\Big]$ are the shares of output per young household available for consumption (output less government expenditure and investment in private capital). Note that H and H^* are positive as the government cannot consume more than total output in the respective country. From Eq. (24) it follows that, ceteris paribus, the scarcer the product the higher the relative price.

Likewise, Eq. (22) can be simplified by using Eq. (23) to:

$$e = -\frac{1}{\rho\mu}\frac{\Phi}{\Phi^*}, \tag{25}$$

with $\Phi \equiv \left(\sigma\,(1-\alpha-(g+u)) - b\left(\sigma\left(q-A^L\right)+A^L\right)\right)\tilde{g}k^{\alpha/(1-\alpha^G)} - A^L k$ and $\Phi^* \equiv \left(\sigma^*\,(1-\alpha^*-(g^*+u^*)) - b^*\left(\sigma^*\left(q^*-A^L\right)+A^L\right)\right)(\alpha/\alpha^*)\,\tilde{g}k^{\eta\alpha^*/(1-\alpha^{*G})} - A^L k^\eta$. The numerator (denominator) represents the net assets position of Home (Foreign). As we are looking at a two-country model, a positive net asset position in one country implies a negative net asset position in the other country, leading to a positive real exchange rate e.

By equating the right-hand side of (25) with the right-hand side of (24) and rearranging the terms, we get: $[(1-\zeta)/\zeta]\,(H/H^*)\,\Phi^* + \Phi = 0$ or:

$$\begin{aligned} F(k) \equiv \left(A^L\right)^{-1}\tfrac{1-\zeta}{\zeta}\tfrac{H}{H^*}\Phi^* + \left(A^L\right)^{-1}\left(\sigma\,(1-\alpha-(g+u)) - b\left(\sigma\left(q-A^L\right)+A^L\right)\right) \\ \tilde{g}k^{\frac{\alpha}{1-\alpha^G}} = k, \end{aligned} \tag{26}$$

which is a function of Home's capital intensity. This implicit equation defines Home's steady state capital intensity, k. Proposition 1 summarizes the conditions under which at least one solution for this equation, and hence a steady state equilibrium, exists.

Proposition 1 Existence of steady state[2]

[2]The proof works very similar to that in 3.2.2.1 in Schelnast (2013). For further details, please consult the authors.

If public expenditure shares are sufficiently small (i.e., $g^* + u^* < 1 - \alpha^*$ and $g + u < 1 - \alpha$), for:

(i) $b = 0$ and $b^* = 0$, one non-trivial steady state exists,
(ii) $0 < b < b^{\max}$ and $0 < b^* < (b^*)^{\max}$, two non-trivial steady states exist.

Proposition 2 Stability of steady state
If public expenditure shares are sufficiently small (i.e., $g^* + u^* < 1 - \alpha^*$ and $g+u < 1-\alpha$), and if Home and Foreign debt ratios are below some upper limits (i.e., $b < b^{\max}$ and $b^* < (b^*)^{\max}$) but non-negative, the higher steady state equilibrium (i.e., where k is larger) is saddle-path stable and the lower steady state is saddle-path unstable.

To determine the stability properties of the economic system, the dynamic equations (17), (20), and (22) have to be analyzed near the steady state equilibria. This can be done by means of the eigenvalues of the Jacobian matrix for the respective steady states. The derivation of the elements of the Jacobian matrix can be found, for example, in Farmer and Schelnast (2013, p. 326f).[3] Extensive numerical simulations reveal that for the lower steady state, two eigenvalues are outside the unit circle, and for the higher steady state, only one steady state is outside the unit circle. Hence, the lower steady state is saddle-path unstable, while the higher steady state is saddle-path stable. Without productive government capital, this property can be proved analytically for the case of equal private capital elasticities in Home and Foreign (see Farmer and Schelnast 2013, p. 315) and zero debt in Home and Foreign (i.e. $b = b^* = 0$) and also for differing private capital elasticities in Home and Foreign (see Farmer and Schelnast 2013, p. 339f).

3 Maximum Sustainable Debt

A debt is called sustainable if a debt to output ratio can be maintained by retaining the pre-determined tax structure.[4] Hence, if the debt ratio is lower than the maximum sustainable debt ratio, no governmental interventions are necessary to prevent an economic collapse. The aim of this section is to define how the maximum sustainable debt ratio is influenced by different economic parameters, such as Foreign's debt ratio, and governmental productive and unproductive expenditures, Home's and Foreign's savings rates, and a change in Home's and Foreign's production elasticities.

As Proposition 1 states, there are two steady state equilibria if debt ratios are sufficiently small. A higher debt ratio shifts the curve $F(k)$ downwards.[5] Hence,

[3]For details on the elements of the Jacobian matrix, please consult the authors.

[4]This assumption is also applied in Rankin and Roffia (2003).

[5]Proved in 3.2.2.1 in Schelnast (2013).

Fig. 1 Maximum sustainable debt

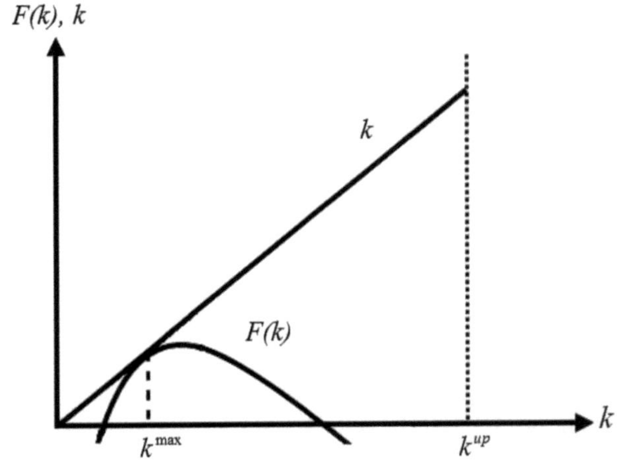

we search for that debt ratio which leads only to a single steady state equilibrium, where the two steady state equilibria coincide. Any higher debt ratio would result in an economic collapse of uncontrolled capital decumulation (=a catastrophe).[6] Therefore, we have to determine the debt ratio that shifts the function $F(k)$ downwards so that it just touches the function k, as shown in Fig. 1.

In this special steady state equilibrium, two mathematical conditions have to hold at the same time, namely the equation for steady state equilibrium, (26), and the slope of $F(k)$ equaling that for k, or equivalently written:

$$F_k'(k) = 1. \tag{27}$$

This provides us with a system of two implicit equations with two unknowns, Home's capital intensity, k, and the maximum sustainable debt ratio, $\overline{b}$. It is not possible to solve this system of equations explicitly. However, we can deduce how a change in an exogenous variable changes the maximum sustainable debt ratio. In particular, we are interested in the effects of the Foreign debt ratio, governmental productive and unproductive expenditures, Home's and Foreign's saving rates, and a change in Home's and Foreign's production elasticities. The analytical derivation of the effects of these economic variables on the limit of Home's government debt can be found in the Appendix. Proposition 3 summarizes the main findings:

Proposition 3 Determinants of maximum sustainable debt limits

If debt ratios are non-negative, $b \geq 0$ and $b^* \geq 0$, Home's maximum sustainable debt limit:

- decreases in relation to Foreign's debt limit, b^*;

[6]If the debt ratio is higher than the maximum sustainable one, the function $F(k)$ never crosses k in the domain $(0, k^{up})$.

- increases in relation to the patience of Home's and Foreign's consumer, σ and σ^*;
- decreases the higher Home's (Foreign's) expenditure share for productive expenses, g (g^*), and unproductive expenses, u (u^*) become;
- decreases in relation to the output elasticity of capital in Home or Foreign, α or α^*.

The higher Foreign's debt ratio, b^*, the lower Home's maximum sustainable debt ratio. This is a typical crowding out effect; a higher supply of foreign debt has to be accompanied by a lower supply of Home's bonds.

If, on the other hand, either Home's or Foreign's consumers become more patient, the maximum sustainable debt ratio increases. If preferences for consumption in the retirement period increase, people are willing to give up today's consumption to increase future consumption. Thus, the demand for savings opportunities increases.

Next, we consider the effect of an increase in Home's government expenditure share on the maximum sustainable debt ratio. The change in maximum sustainable debt caused by an increase in Home's government productive expenditures is given by[7]:

$$\frac{d\bar{b}}{dg} = \frac{(\phi/H) - \sigma}{(1-\sigma)A^L + \sigma q} = \frac{-\left[\left(\sigma q + (1-\sigma)A^L\right)/(\alpha H)\right]k(\alpha + bq)}{(1-\sigma)A^L + \sigma q} < 0. \quad (28)$$

Equation (28) is negative, meaning that the higher the expenditure share, the lower the maximum sustainable debt ratio. Equation (28) is negative as debt ratios are assumed to be non-negative. Likewise, the maximum sustainable debt ratio is decreasing in Foreign's government productive expenditure ratio.

Surprisingly, a change in Home's government unproductive expenditures has exactly the same effect on the change in Home's maximum sustainable debt ratio as a change in productive expenditures. Hence, whether the government increases its productive or unproductive expenditures has the same consequence for the maximum sustainable debt ratio. However, the capital intensity in the steady state equilibrium with the maximum sustainable debt ratio is higher in the case of an increase in productive expenditures compared to an increase in unproductive expenditures.[8]

If the output elasticity of capital in Home or Foreign, α or α^*, increases, the maximum sustainable debt ratio decreases. Thus, the maximum sustainable debt ratio declines in the case of technological progress by means of which capital becomes more productive compared to labor either in Home or Foreign.

---

[7]The basis for this derivative is given in the Appendix.

[8]The proof is given in the Appendix.

4 Conclusion

We have shown that the relationship between many economic variables and maximum government debt ratios is surprisingly clear. High foreign debt ratios have a negative impact on the home country's maximum debt limit. As one might expect, the relationship between private saving rates and maximum debt is positive. This can explain the prevailing high government debt ratios in Japan as savings rates in the Asian region are traditionally high. Both productive and unproductive government expenditures influence government debt limits negatively and to the same extent. However, steady state capital intensity and welfare levels differ. An increase in the output elasticity of capital reduces the possible debt ratio of the home country.

Finally, we wish to thank Karl Farmer for his support in our years of study. We appreciate his research on OLG models and debt dynamics, which inspired us to capture these issues and cover them within our diploma and doctoral theses.

Appendix

We seek the debt ratio, $\overline{b}$, for which only one steady state equilibrium exists. This debt ratio is called the maximum sustainable debt ratio as for a higher debt ratio there is no steady state equilibrium. Both, Eqs. (26) and (27) have to be fulfilled simultaneously. Through rearranging Eqs. (26) and (27), we obtain:

$$S^1\left(k,\overline{b}\right) \equiv A^L F\left(k,\overline{b}\right) - A^L k = \frac{1-\zeta}{\zeta}\frac{H}{H^*}\Phi^* + \Phi = 0 \tag{29}$$

and

$$S^2\left(k,\overline{b}\right) \equiv F_k'\left(k,\overline{b}\right) - 1 = 0. \tag{30}$$

Equation (30) equals:

$$\left[\frac{(1-\zeta)}{\zeta}H\left(A^L - \frac{\sigma^* b^* q^2}{\alpha^*}\right)k^\eta + H^* k\left(A^L - \frac{\sigma\overline{b}q^2}{\alpha}\right)\right] = \frac{\Phi}{H}A^L\left(kH^* - k^\eta H\right), \tag{31}$$

which can be derived by calculating the partial derivatives $F_k'(k,b)$ and using Eq. (29) to simplify terms.

Equations (29) and (30) represent a system of two equations with two unknowns, k and $\overline{b}$. We are interested in the change in the maximum sustainable debt ratio if a parameter z, $z \in \{b^*, \sigma, \sigma^*, \alpha, \alpha^*, g, g^*, u, u^*\}$, changes. Thus, by totally differentiating Eqs. (29) and (30), we obtain:

$$\begin{bmatrix} \frac{\partial S^1}{\partial k} & \frac{\partial S^1}{\partial \overline{b}} \\ \frac{\partial S^2}{\partial k} & \frac{\partial S^2}{\partial \overline{b}} \end{bmatrix} \begin{bmatrix} \frac{\partial k}{\partial z} \\ \frac{\partial \overline{b}}{\partial z} \end{bmatrix} = \begin{bmatrix} -\frac{\partial S^1}{\partial z} \\ -\frac{\partial S^2}{\partial z} \end{bmatrix}. \tag{32}$$

Note that $\partial S^1/\partial k = A^L\left(F'_k - 1\right) = 0$ because of Eq. (30). By using this fact and Cramer's rule, the following equation then follows from Eq. (32):

$$\frac{\partial \overline{b}}{\partial z} = -\frac{\partial S^1}{\partial z} / \frac{\partial S^1}{\partial \overline{b}}. \tag{33}$$

The denominator of Eq. (33) is negative and given by:

$$\frac{\partial S^1}{\partial \overline{b}} = (-1)\left(\sigma q + (1-\sigma) A^L\right) \tilde{g} k^{\frac{\alpha}{1-\alpha^G}} < 0. \tag{34}$$

Hence, whether the maximum sustainable debt ratio decreases or increases depends on whether $\partial S^1/\partial z$ is smaller or larger than zero.

First, we consider the effect on the maximum sustainable debt ratio if Foreign's debt ratio increases. Thus, we have to calculate the partial derivative $\partial S^1/\partial b^*$, which is negative:

$$\frac{\partial S^1}{\partial b^*} = \frac{1-\zeta}{\zeta}\frac{H}{H^*}(-1)\left(\left(\sigma^* q + \left(1-\sigma^*\right) A^L\right) \frac{\alpha}{\alpha^*} \tilde{g} k^{\eta \frac{\alpha}{1-\alpha^G}}\right) < 0. \tag{35}$$

Next, we consider the effects of a change in the time preference of Home's consumption: an increase in β. If β increases, σ also increases and vice versa.

$$\frac{\partial S^1}{\partial \sigma} = \left((1-\alpha-(g+u)) - b\left(q - A^L\right)\right) \tilde{g} k^{\frac{\alpha}{1-\alpha^G}} > 0 \tag{36}$$

The inequality in (36) follows from the condition that the tax rate is smaller than one. The government cannot collect more money than the workers earn. The tax rate is given by Eq. (21) and can be simplified to: $\tau = (1-\alpha)^{-1}\left[b\left(q - A^L\right) + (g+u)\right]$ in the steady state equilibrium. From $\tau < 1$, through rearranging terms, we obtain: $(1-\alpha-(g+u)) - b\left(q - A^L\right) > 0$.

Similar argumentation leads to the following inequality:

$$\frac{\partial S^1}{\partial \sigma^*} = \frac{1-\zeta}{\zeta}\frac{H}{H^*}\left(\left(1-\alpha^* - \left(g^* + u^*\right)\right) - b^*\left(q^* - A^L\right)\right) \frac{\alpha}{\alpha^*} \tilde{g} k^{\eta \frac{\alpha}{1-\alpha^G}} > 0. \tag{37}$$

Next, we are interested in how a change in g influences the maximum sustainable debt. Thus, we need to calculate the following partial derivative:

$$\frac{\partial S^1}{\partial g} = \left[\frac{(1-\zeta)}{\zeta} \frac{H}{H^*} \Phi^{*\prime}_g + \frac{(1-\zeta)}{\zeta} \frac{H'_g H^* - H H^{*\prime}_g}{(H^*)^2} \Phi^* + \Phi'_g \right]. \tag{38}$$

This can be rearranged by using Eq. (29) to:

$$\begin{aligned} \tfrac{\partial S^1}{\partial g} &= \tfrac{1}{H^*} \tfrac{\alpha^G}{1-\alpha^G} \tfrac{1}{g} \left(\tfrac{(1-\zeta)}{\zeta} H \left(A^L - \tfrac{\sigma^*}{\alpha^*} b^* q^2 \right) k^\eta + H^* \left(A^L - \tfrac{\sigma}{\alpha} b q^2 \right) k \right. \\ &\left. - \tfrac{\Phi}{H} A^L \left[k H^* - k^\eta H \right] \right) + \left(\tfrac{\Phi}{H} - \sigma \right) \tfrac{1}{\alpha} q k. \end{aligned} \tag{39}$$

Expression (39) can be simplified using Eqs. (29) and (31):

$$\frac{\partial S^1}{\partial g} = \frac{1}{\alpha} q k \left(\frac{\phi}{H} - \sigma \right). \tag{40}$$

Hence, the change in maximum sustainable debt caused by a change in g is given by:

$$\frac{\partial \bar{b}}{\partial g} = -\frac{\partial S^1}{\partial g} \Big/ \frac{\partial S^1}{\partial b} = -\frac{\alpha^{-1} q k \left((\phi/H) - \sigma \right)}{(-1) \left(\sigma q + (1-\sigma) A^L \right) q k \alpha^{-1}} = \frac{\left((\phi/H) - \sigma \right)}{\left(\sigma q + (1-\sigma) A^L \right)}. \tag{41}$$

Differentiating Eq. (29) with respect to g^* und using a similar simplification approach leads to:

$$\begin{aligned} \tfrac{\partial S^1}{\partial g^*} &= \tfrac{1-\zeta}{\zeta} \tfrac{1}{\alpha^*} q k^\eta \tfrac{1}{H^*} \left(\tfrac{\phi^*}{H^*} - \sigma^* \right) = \\ &= \tfrac{1-\zeta}{\zeta} \tfrac{1}{\alpha^*} q k^\eta \tfrac{1}{H^{*2}} \left(- \left[\tfrac{\sigma^* q + (1-\sigma^*) A^L}{\alpha^*} \right] k^\eta \left(\alpha^* + b^* q \right) \right) < 0. \end{aligned} \tag{42}$$

The inequality holds for all non-negative debt ratios, b^*.

Likewise, the partial derivative of S^1 with respect to u and u^* is given by:

$$\begin{aligned} \frac{\partial S^1}{\partial u} &= \left[\tfrac{(1-\zeta)}{\zeta} \tfrac{H}{H^*} \Phi^{*\prime}_u + \tfrac{(1-\zeta)}{\zeta} \tfrac{H'_u H^* - H H^{*\prime}_u}{(H^*)^2} \Phi^* + \Phi'_u \right] = \left[\tfrac{\Phi}{H} - \sigma \right] \\ &\tilde{g} k^{\alpha/(1-\alpha^G)} \end{aligned} \tag{43}$$

and

$$\begin{aligned} \tfrac{\partial S^1}{\partial u^*} &= \left[\tfrac{(1-\zeta)}{\zeta} \tfrac{H}{H^*} \Phi^{*\prime}_{u*} + \tfrac{(1-\zeta)}{\zeta} \tfrac{H'_{u*} H^* - H H^{*\prime}_{u*}}{(H^*)^2} \Phi^* + \Phi'_{u*} \right] \\ &= \tfrac{(1-\zeta)}{\zeta} \tfrac{H}{H^*} \left[\tfrac{\Phi^*}{H^*} - \sigma^* \right] \tfrac{\alpha}{\alpha^*} \tilde{g} k^{\eta \alpha^* / (1-\alpha^{*G})}. \end{aligned} \tag{44}$$

Next, we are interested in the change in maximum sustainable debt ratio if the production elasticity changes. Thus, we have to calculate the derivative of S^1 with respect to α:

$$\frac{\partial S^1}{\partial \alpha} = \left[\frac{(1-\xi)}{\xi}\frac{H}{H^*}\Phi^{*\prime}{}_\alpha + \frac{(1-\xi)}{\xi}\frac{H'_\alpha H^* - HH^{*\prime}{}_\alpha}{(H^*)^2}\Phi^* + \Phi'_\alpha\right] = \frac{1}{\alpha}\frac{1}{H}\left[\Phi A^L - HA^L - H\sigma q\right]k$$
$$= -\frac{1}{\alpha H}\frac{qk}{\alpha}\left[\left(\sigma q + (1-\sigma)A^L\right)\left((1-(g+u)) + bA^L\right)\right]k < 0 \tag{45}$$

The second equality follows from calculating the partial derivatives with respect to α and simplifying the term using the functional relations given in (29) and (31). This partial derivative is negative for non-negative debt ratios in Home, $b \geq 0$. Hence, the change in the maximum sustainable debt ratio caused by a change in the production elasticity is negative, $\partial\overline{b}/\partial\alpha = -\left(\partial S^1/\partial\alpha\right)/\left(\partial S^1/\partial b\right)$ as both denominator and nominator are negative.

A similar approach can be used to calculate a change in the maximum sustainable debt ratio caused by a change in Foreign's production elasticity:

$$\frac{\partial S^1}{\partial \alpha^*} = -\frac{(1-\xi)}{\xi}\frac{H}{H^{*2}}\frac{1}{\alpha^*}\frac{qk^\eta}{\alpha^*}\left[\left((1-(g^*+u^*)) + b^*A^L\right)\left(\sigma^* q + (1-\sigma^*)A^L\right)\right] k^\eta < 0, \tag{46}$$

where the inequality follows from a non-negative debt ratio in Foreign.

Finally, we have to show that the effects on the capital intensity are different depending on whether the government increases productive or unproductive expenditures. By using Cramer's rule as well as (29), (31), (41), and (43), we obtain:

$$\frac{\partial k}{\partial g} - \frac{\partial k}{\partial u} = \frac{\partial S^2/\partial g - \partial S^2/\partial u}{-\partial S^2/\partial k} = \frac{\alpha^G k}{\left(1-\alpha^G-\alpha\right)g} > 0. \tag{47}$$

Thus, in the case that the maximum sustainable debt ratio is reached, capital intensity is higher if the government invests in productive expenditures.

References

De la Croix D, Michel P (2002) A theory of economic growth. Dynamics and policy in overlapping generations. Cambridge University Press, Cambridge

Diamond PA (1965) National debt in a neoclassical growth model. Am Econ Rev 55:1135–1150

Farmer K (2006) Reducing public debt under dynamic efficiency: transitional dynamics in Diamond's OLG model. Atl Econ J 34:195–208

Farmer K, Schelnast M (2013) Growth and international trade. An introduction to the overlapping generations approach. Springer, Berlin, Heidelberg

Farmer K, Zotti J (2010) Sustainable government debt in a two-good, two-country overlapping generations model. Int Rev Econ 57:289–316

Rankin N, Roffia B (2003) Maximum sustainable government debt in the overlapping generations model. Manch Sch 71:217–241

Schelnast M (2013) Public debt: maximum sustainability and welfare effects of debt reduction in a two-country two-good diamond-type OLG model. Dissertation, University of Graz

The Long Italian Stagnation and the Welfare Effects of Outsourcing

Jacopo Zotti

1 Introduction

In the early nineties, the Italian economy was one of the world's largest per GDP. This successful performance was the outcome of a long period of sustained growth, which had started immediately after the end of WW2 and continued until the late eighties. In 1991, per-capita GDP was above EU average. Starting in that year however, the so-called "Italian Economic Miracle" (Nardozzi 2003) seemed to vanish quickly, and a long period of stagnation took its place. Italy started to diverge from its major EU partners and in terms of real per-capita GDP, in 2014 it was back to the levels as of 1997 (IMF 2015). This long period of internal stagnation almost coincides with a phase of extraordinarily intense globalization. Like all other developed countries, Italy took active part in this process, but, contrary to its partners, seemed unable to benefit from it.

The fact that a country may fail to benefit from greater openness comes quite at odds with the conventional economic wisdom that indeed predicts net benefits from increased integration. However, this perception is quite widespread among Italian scholars (e.g. Trento 2003; Ciocca 2004; Accetturo et al. 2013) who believe that Italy failed to gain from globalization or even lost from it because of a number of structural weaknesses. The broad consensus around this position, however promptly clashes with the heterogeneity of ideas around which features of the economy really prevented the country from taking advantage of globalization. The debate is particularly intricate because most positions rely on qualitative empirical analysis and deductive reasoning rather than on quantitative methods. For this reason it

J. Zotti (✉)
Fondazione Eni Enrico Mattei, Venice, Italy

Department of Political and Social Sciences, University of Trieste, Trieste, Italy
e-mail: Jacopo.Zotti@econ.units.it

B. Bednar-Friedl, J. Kleinert (eds.), *Dynamic Approaches to Global Economic Challenges*, DOI 10.1007/978-3-319-23324-6_6

is quite difficult to identify clear causational linkages among the different factors considered and also a hierarchy among them is not immediately clear. In an attempt to identify the most relevant weaknesses according to the ongoing debate, a main issue seems to be the generalized low propensity towards innovation. According to some authors like for example Faini and Sapir (2005) and Ciocca (2010), this might be a basic reason for the slow pace of adoption of the new information and communication technologies (ICT) by Italian firms. Vaciago (2003) and Accetturo et al. (2013) recognize ICT as a crucial factor of competitiveness for national firms during these decades of intense globalization while other authors (e.g. Pagano and Schivardi 2003; Rossi 2004) believe the relatively small size of the average Italian firm to be a major impediment to the slow adoption of ICT. The scarce R&D efforts seem a valid explanation also for the structure of the Italian specialization pattern, which a large body of literature (see for example Ciocca 2004; D'Ippoliti and Roncaglia 2011) retains excessively biased towards traditional and low-tech productions (typically: textile products, apparels, shoes, furniture, hydraulics and non-metal manufacturing). In this perspective, Italy might have suffered from globalization mainly because of the involvement of the emerging countries, which are characterized by similar specialization patterns, and enjoy huge cost advantages relatively to Italy.

In this chapter, we share the view that Italy might have effectively failed to gain from globalization, or rather to have suffered from it. However, we focus on another salient aspect of the economy as a possible explanation for such failure, namely the scarce degree of competition on domestic product markets.[1] To lend support to this view, we provide a highly stylized model of a small open economy with Cournot-oligopolistic markets and foreign outsourcing. We assume two sectors, manufactures and services and we mimic the higher degree of competition in the goods' markets relatively to the service markets by assuming the former to be perfectly competitive and the latter Cournot-oligopolistic. We approximate the level of economic integration of Italy in the world economy with an exogenous tariff on intermediates.[2] The economy altogether is thus subject to two distortions, i.e. the number of oligopolists in the service market and the tariff on intermediates.

The interaction between these two distortions constitutes a typical second-best framework (Lipsey and Lancaster 1956), which allows us to prove that a lower tariff rate, leading to a more intensive degree of outsourcing is not necessarily beneficial for the economy if internal markets are (even partly) overregulated. For a given level of competition in the oligopolistic service market, consumer welfare is an

[1]On this point, a large economic literature (e.g. Barca 1997; Faini 2003; Faini et al. 2005; Nardozzi 2004; Ciocca 2007; Forni et al. 2010) maintains that markets in Italy were and still are less competitive than in most OECD countries. Bianco et al. (2012), for example, provide evidence of a stable or an even growing Lerner index on several final product markets throughout the whole nineties. The need for more competitive markets is also a primary policy issue (OECD 2005; CNEL 2007; Christopoulou and Vermeulen 2008) and a major objective of the National Reforms' Program by the Italian Ministry of Economy and Finance (MEF 2011).

[2]The average tariff for Italy was decreasing in the period 1990–2010 (Accetturo et al. 2013).

inverted U-shaped function in the level of tariffs and an optimal tariff does exist. When competition in the oligopolistic sector is scarce, a sufficiently low tariff on intermediates may induce a welfare loss. Relatively to the oligopolistic sector, the competitive sector overproduces, and the marginal welfare in this sector is lower than in the competitive. Similarly, when outsourcing is subject to tariffs, oligopoly and *not* perfect competition is the desirable market regime, and an optimal number of firms in the oligopolistic sector can be determined. A tight market regulation leads to a reduction of production in the oligopolistic sector and more resources become available for the production of the competitive good, and this may generate a welfare gain. More importantly, the optimal number of oligopolists is inversely related to the level of tariffs. When economic integration proceeds, the domestic competition policy should react and become stricter, otherwise the greater openness translates into aggregate welfare losses. From this perspective, these findings lend analytical support to the idea that Italy may have effectively lost from globalization.

The model presented in this chapter belongs to a quite recent line of research on Cournot oligopoly in general equilibrium, originally initiated by Neary (2003). An overview of this literature is contained in Zotti and Lucke (2014) who depart from the one-country-one-factor structure of Crettez and Fagart (2009) to study the welfare optimality of trade and competition policies in small open oligopolistic economies (SOOE) with trade in final goods. We extend the small open oligopolistic economy framework of Zotti and Lucke (2014) to incorporate a rudimental form of trade in intermediates, which is the prominent feature of the current wave of globalization (see for Italy the study by Breda and Cappariello 2012). At the same time we maintain the original static structure of the model in consideration of the substantial lack of growth over the last two decades in Italy. Moreover, we approximate the relative closedness of the Italian service sector[3] in comparison to the good sector with the assumption that Italy does not trade services.[4] Under the assumption of balanced trade, the economy imports intermediates and exports manufactures.

The second section of this chapter seeks to provide an overview of the main features of the Italian economy over the last two decades. Preliminarily, it includes some basic empirical evidence of the stagnation. The third section describes the structure of the model of a small open oligopolistic economy (SOOE) with outsourcing, and the fourth section derives results about globalization and welfare-optimal competition policy. The fifth section draws some conclusions.

[3]Italian trade in manufactures varies around 80–90 % of the trade balance (see Amighini and Chiarlone 2004; Accetturo et al. 2013).

[4]From the technical point of view of the modeling structure, this assumption does not impinge on the results.

2 Economic Stagnation and Inability of Gaining from Globalization

The features of the Italian economy in the last two decades are the object of two contiguous strands of literature. These are the debate on the reasons for the economic stagnation and the historical discussion on the structural weaknesses of the economy, which dates back to the times of the national unification. Most of the literature on the stagnation considers various aspects of the economy, and uses deductive reasoning (Rossi 2004, p. 640) supported by qualitative data observation to provide intuitions concerning their role in the crisis. Globally, this literature indicates many reasons for the observed stagnation, and from this perspective, it complements and updates the older debate on the structural weaknesses of the economy.[5] Altogether, these two strands of literature are relevant for the debate on Italy's inability to gain from globalization because of the comprehensive overview they provide on the alleged weaknesses of the economy. From a methodological point of view, in fact, this debate is very similar to the literature on the economic stagnation, as it tries to infer causational linkages between a given feature of the economy and the missed benefits from globalization.

Based on these nonetheless distinct strains of literature, this section seeks to provide a broad overview of the main features of the Italian economy, and to highlight which of them seem to have impeded Italy from gaining from globalization. An overview of these features is given in Table 1. The table records the main contributions of the literature on the economic stagnation, and reveals the deep heterogeneity of the debate. In this overview, we focus on the major weaknesses of the economy and we seek to deal with them along a unified line of reasoning, which starts from the central role of the stagnant TFP and proceeds by searching for its possible determinants. Table 1 complements the overview, as it includes those aspects excluded from this discussion.

We document the Italian economic stagnation by comparing the evolution of the real per-capita GDP between Italy and other major developed economies. Figure 1 shows the Italian GDP as a fraction of the GDP of the EU-14 (i.e. the EU prior to the Fifth Enlargement, excluding Italy) for the period 1951–2008. The process of economic convergence prescribed by neoclassical growth theory is clearly observable from the end of WW2 to the beginning of the nineties. Since then, fully completed convergence turned into lengthy divergence. In order to illustrate the severity of the Italian stagnation, we present (Fig. 2) the overall performance of the economy in terms of real per-capita GDP growth rates for the period 1951–2008. Note that it is sensible to analyse the Italian growth performance after studying the international comparison, since this allows excluding any neoclassical-type convergence process as a major source of the slowdown.

[5] An example of this complementarity is Faini (2003) who includes the historical north–south divide as an explanation for the Italian stagnation of the last two decades.

Table 1 Features and weaknesses of the Italian economy according to the relevant literature on the stagnation

Is this feature a major weakness of the Italian economy?	[1]	[2]	[3]	[4]	[5]	[6]	[7]	[8]	[9]	[10]
Labor productivity slowdown	0	Y	Y	0	Y	Y	Y	Y	Y	Y
TFP slowdown	0	Y	Y	0	Y	Y	Y	Y	0	Y
Decline in the labor input	0	N	0	Y	N	0	0	0	0	0
Insufficient R&D activity	N	0	Y	Y	Y	Y	Y	Y	0	Y
Slowdown in capital accumulation	0	N	0	0	N	N	0	0	Y	0
Delay in ICT incorporation	Y	0	Y	0	Y	Y	0	0	0	Y
Tax evasion	0	0	Y	0	0	0	0	Y	0	0
Corruption	0	0	0	0	0	0	0	0	0	Y
Inadequate specialization pattern	0	N	Y	0	Y	Y	Y	0	Y	0
Low endowment of human capital	0	Y	Y	Y	N	0	Y	0	0	0
Lack of competition on markets	0	Y	0	0	Y	Y	0	Y	0	Y
Labor market rigidities	0	0	0	0	0	0	0	0	0	N
Labor costs	0	0	0	0	0	N	0	N	0	N
Labor market reforms	0	0	0	0	0	0	0	Y	Y	0
Poorly functioning financial markets	0	0	0	0	N	0	0	0	0	Y
Public debt burden	0	0	0	Y	Y	0	0	Y	0	0
Stock and quality of physical infrastructure	0	0	0	Y	Y	0	0	Y	0	Y
Stock and quality material infrastructure	0	0	0	Y	Y	0	0	Y	0	0
Insufficient firms' size	N	0	Y	Y	Y	Y	0	Y	0	Y
Biased income redistribution	0	0	0	0	Y	0	0	0	Y	0
Aggregate demand weakness	0	0	0	0	0	N	0	0	Y	0
Italy's north–south divide	0	Y	0	N	Y	0	0	0	0	0
Inflation	0	0	0	0	N	0	0	0	Y	0

Authors: [1]: Vaciago (2003); [2]: Faini (2003); [3]: Trento (2003); [4]: Toniolo (2004); [5]: Ciocca (2004); [6]: Rossi (2004); [7]: Faini and Sapir (2005); [8]: Ciocca (2010); [9]: D'Ippoliti and Roncaglia (2011); [10]: Accetturo et al. (2013)
Abbreviations: *Y* yes, *N* no, *0* irrelevant

A basic feature of the Italian stagnation is undoubtedly a poor TFP dynamics. This is widely recognized by numerous studies which use different methods to show the unambiguous role of the stagnant TFP in the slowdown of the labor productivity. Faini (2003) and Ciocca (2004) study the data on capital accumulation and conclude that this remained substantially constant in the nineties. Daveri and Jona Lasinio (2006) perform a decomposition of labor productivity growth and show that falling labor productivity and not labor input is the reason for the observed decline in real per-capita income growth. Noticeably, these results are fully confirmed almost a decade later by Orsi and Turino (2014) who apply the business cycle accounting

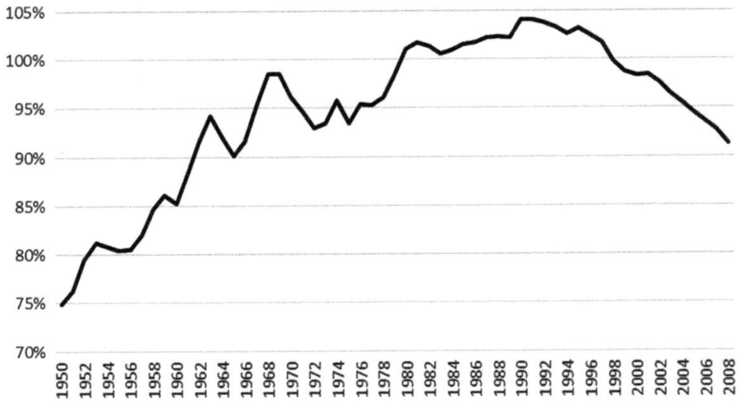

Fig. 1 Italy's real per-capita GDP as a fraction of EU-14 average (*Source*: Own calculations from Groningen Growth and Development Centre data, www.ggdc.net)

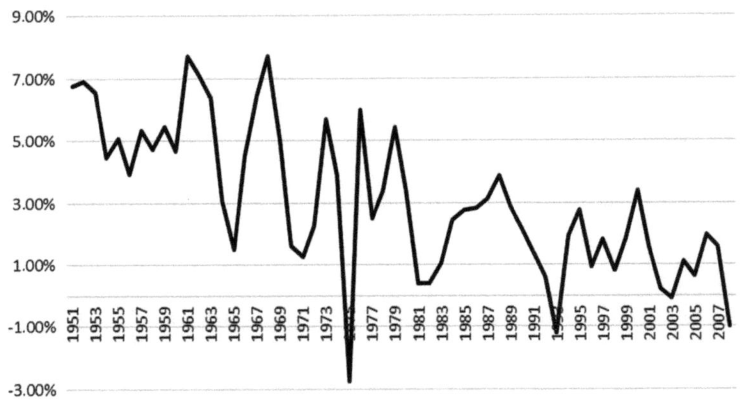

Fig. 2 Italy's real per-capita GDP growth rates (*Source*: Own calculations from Groningen Growth and Development Centre data, www.ggdc.net)

procedure by Chari et al. (2007) and show that the labor input actually improved considerably starting in the mid-nineties.[6]

There are two main explanations for the poor TFP dynamics, which are both taken from the literature on the sources of the EU–US productivity divide since the mid-nineties. These are the labor market reforms of the nineties (e.g. Blanchard and Landier 2002; Dew-Becker and Gordon 2012) and the insufficiency of investments in R&D and ICT (e.g. van Ark et al. 2008). In the case of Italy, there is in fact robust

[6]Further studies with similar results are Daveri (2002), Brandolini and Cipollone (2003), Daveri (2004), and Fachin and Gavosto (2010).

evidence both for the trade-off between employment and productivity[7] and for the direct effect of the insufficient level of R&D on the TFP (e.g. Parisi et al. 2006; Fachin and Gavosto 2010). The level of R&D expenditures as a share of GDP has been constantly below the EU-average (OECD 2006, 2012) with R&D intensity in the private sector far lower than in the other EU countries. During the nineties, in particular, the business R&D efforts in the nineties showed a net drop in comparison to the previous decade. According to Venturini (2004), also the pace of investment in ICT followed a similar path in Italy.[8] Bassanetti et al. (2004) perform a growth accounting exercise aimed at measuring ICT contribution to growth, and their results show a negligible impact of ICT on the TFP dynamics. The slow adoption of ICT by Italian firms is one of the factors hampering Italy's exploitation of the opportunities offered by globalization according to a conspicuous body of literature (e.g. Vaciago 2003; Rossi 2004; Ciocca 2004 and more recently, Accetturo et al. 2013).

The inadequate intensity of business R&D and the delayed adoption of ICT have motivated an extensive literature aiming at exploring their main determinants. Two explanations for this generalized inertia towards innovation are worth mentioning here. One position that enjoys a broad consensus in the debate (e.g. Trento 2003; Toniolo 2004; Rossi 2004), points to the average size of the Italian firm, which international comparison reveals to be smaller than in other major partners. According to these authors, there exists a dimensional threshold, below which a single firm faces serious constraints in engaging successful research activities, or even in adopting (relatively) costly modern technologies.[9] Regarding the latter measure, several authors (e.g. Faini and Sapir 2005) consider the simple scarcity of human capital in a company as a major obstacle for the adoption of the modern ICT. A second explanation focuses on the pattern of specialization of the economy, which is the object of a huge body of literature starting in the sixties[10] and provides empirical evidence that Italy has a comparative advantage in traditional (low-tech) sectors. According to this view, these sectors have an intrinsically low propensity to innovate.

From the perspective of the literature on the effects of globalization on Italy, the biased structure of the economy provides a sensible explanation for the negative consequences from increased integration. Italy in fact may have suffered from globalization because its competitiveness has been challenged by the emergence

[7]Papers that, with different approaches, confirm the employment-productivity trade-off for Italy are Boeri and Garibaldi (2007), Lucidi and Kleinknecht (2010), Lucidi (2012), Jona Lasinio and Vallanti (2011) and, more recently, Orsi and Turino (2014).

[8]Pilat et al. (2002) however distinguish between "fast-adopters" (UK, Netherlands, Sweden and Finland) and "laggards" (Italy and Spain and, to some extent, Germany and France).

[9]The debate on the reasons for the inadequately small size of Italian firms is very wide. Some insights can be found foremost in Onida (2004) and in Trento (2003) as well as in Ciocca (2004) and Accetturo et al. (2013).

[10]For a review of this literature, see for example, Amighini and Chiarlone (2004) and Federico and Wolf (2012) for a more historical perspective.

of the developing countries in the international arena.[11] Among the first to believe that the specialization pattern has become "inadequate" are Onida (1999) and Trento (2003). Faini and Sapir (2005) support this view by calculating the Balassa index of Revealed Comparative Advantage for Italy and by studying its evolution through time. The size of the literature studying the Italian specialization pattern contrasts however with the tiny number of papers searching for the reasons behind it. The explanations provided are a general scarcity of human capital (e.g. Faini 2003; Faini and Sapir 2005; Boeri et al. 2005), and, again the insufficient average size of Italian firms. Several authors (e.g. Trento 2003; Onida 2004; Ciocca 2004) argue in fact that a change in the specialization pattern is generally less probable for smaller firms. An even greater shortcoming of this literature however, seems to be the inability to demonstrate any causation between the proven biases in the specialization pattern and Italy's difficulties on world markets during this wave of globalization. We are not aware in fact of any paper proving this causation through quantitative methods.

The two explanations for the low propensity to innovate are seriously challenged by Sterlacchini and Venturini (2014) and, from a quite different perspective, by Ciocca (2010). The former authors estimate the long-term elasticity of TFP with respect to the level of R&D (measured as the share of R&D expenditures on value added) for a set of five OECD countries including Italy. In the case of Spain, which shares a similar industrial structure with Italy both for firms' size and specialization pattern, the estimated elasticity is 0.19 while for Italy it amounts to 0.08–0.12 (with France at 0.19–0.21). Noticeably, the authors observe that "in the typical research-intensive industries, Italian firms devote to R&D half of the share of value added invested in the most industrialized countries" (Sterlacchini and Venturini 2014, p. 193). The authors propose indeed a different explanation for the insufficient R&D investments, which points to the general tightness of financial constraints at firm level. In their view, these are a direct consequence of the structure of the Italian banking system (in this vein see also Accetturo et al. 2013).

Ciocca (2010) suggests that firms quickly lost their propensity to innovate following a series of policies, which deeply changed the domestic business environment. On foreign markets, the competitiveness of Italian firms was inflated by the undervaluation of the lira, which started with the strong depreciation in the early nineties (Italy abandoned the European Exchange Rate Mechanism as of September 1992) and persisted until 2002. Internally, generous public spending, wage moderation (which started with the July-1993 Tripartite agreement among government, business organizations and trade unions) and a deliberately weak competition policy gave a major contribution to soaring company profits. Widespread and rapidly rising tax evasion further explains the weakened propensity of firms to engage in innovation.

[11]Against this view, Fortis and Curzio (2003) believe that the main threat for the Italian manufacturing is due to the "asymmetric" (i.e. unfair and illegal) competition by China.

3 The Model

We denote the two sectors of the economy by X and Y, where the former is competitive and the latter oligopolistic. Production of X requires value added, which is produced using labor L and capital K, and a foreign intermediate O. The use of input O reflects the delocalization choice of domestic firms. The tariff rate on imported intermediates is equal to a percentage τ of their price. For simplicity, tariffs take the form of a monetary transfer to consumers. Production of Y requires only labor and capital, which are available in fixed supply at $\overline{L}$ and $\overline{K}$. Primary production factors are fully mobile between sectors, but immobile internationally. The model is static, so that investment is zero. Hence, domestic demand includes solely final consumption. In the case of X, it is necessary to distinguish between domestic supply X^S and demand X^D, where the surplus is exported and export proceeds are used to finance imports O, i.e. foreign trade is always balanced.

3.1 Households

The economy is populated by $\overline{L}$ homogeneous private agents. Their preferences are described by a standard Cobb–Douglas utility function:

$$U(X, Y) = X^{\varphi} \cdot Y^{1-\varphi}, 0 < \varphi < 1 \tag{1}$$

Agents are endowed with one unit of labor each, which they supply inelastically at the nominal wage W. In addition, they lend private nominal wealth $P_K\overline{K}$ at the rental rate r to firms, which use the physical capital stock $\overline{K}$ for production. Private agents are price takers in both factor markets. Monetary private income, I, consists of primary factor income, tariffs and profits in the oligopolistic sector:

$$I = W \cdot \overline{L} + rP_K \cdot \overline{K} + E \cdot \tau \overline{P}_O \cdot O + \Pi^Y \tag{2}$$

Here E is the nominal exchange rate, $\overline{P}_O$ is the world price of the imported intermediate O and Π^Y are the monetary profits of the oligopolistic sector Y. Utility (1) is maximized under the following budget constraint:

$$\overline{P}_X \cdot X + P_Y \cdot Y = I \tag{3}$$

where $\overline{P}_X$ is the world market price of commodity X and P_Y the price of commodity Y. Both prices are expressed in home currency. Utility maximizing quantities are

$$X = \varphi \cdot \frac{I}{\overline{P}_X} \tag{4}$$

$$Y = (1 - \varphi) \cdot \frac{I}{P_Y} \tag{5}$$

Note that demand (5) excludes monopoly in sector *Y*, i.e. $N \neq 1$, because price elasticity is one.

3.2 Firms

Firms in sector *X* employ value added *V* and intermediate *O* according to a Cobb–Douglas technology with constant returns to scale:

$$X^S = A^X \cdot V^\eta O^{1-\eta} \tag{6}$$

where $V = A^V \left(K^V\right)^\alpha \left(L^V\right)^{1-\alpha}$. The optimal quantity of value added is

$$V = \frac{X^S}{A^X}\left[\frac{\eta}{1-\eta}\frac{(1+\tau)\overline{P}_O}{P_{V,x}}\right]^{1-\eta} \tag{7}$$

with $P_{V,x} = \frac{1}{A^V}\left(\frac{rP_K}{\alpha}\right)^\alpha\left(\frac{W}{1-\alpha}\right)^{1-\alpha}$, while the optimal intermediate demand is

$$O = \frac{X^S}{A^X}\left[\frac{1-\eta}{\eta}\frac{P_{V,x}}{(1+\tau)\overline{P}_O}\right]^{\eta} \tag{8}$$

where the assumption $\tau \geq 0$ is sufficient for a positive demand.

In sector *Y*, output is produced using only value added with a Cobb–Douglas technology where total factor productivity is A^Y and the capital production elasticity is β. Within each sector, firms are completely homogeneous. Sector *X* is perfectly competitive and many atomistic firms produce and sell their output at world prices. In sector *Y* only few and relatively large business units are active, which operate only on domestic markets and behave strategically as Cournot oligopolists. Despite their non-atomistic dimension, they remain relatively small with respect to the whole economy, i.e. they do not enjoy monopsony power. As argued by Neary (2003), this is crucial, as only through this assumption are single actors prevented from influencing macroeconomic variables so that Cournot oligopoly can be modeled rigorously in general equilibrium.

The total number *N* of oligopolistic firms is exogenous. Since firms are fully identical, sectoral inputs and output are

$$K^Y = N \cdot K_i^Y, \qquad i = 1, 2, \ldots N \tag{9}$$

$$L^Y = N \cdot L_i^Y, \qquad i = 1, 2, \ldots N \tag{10}$$

$$Y = N \cdot Y_i, \qquad i = 1, 2, \ldots N \tag{11}$$

Due to constant returns to scale, cost minimization yields linear cost functions in both sectors:

$$C^X\left(X^S\right) = m_x \cdot X^S \tag{12}$$

$$C^Y(Y) = m_y \cdot Y \tag{13}$$

where $m_x = \frac{1}{A^X}\left(\frac{P_{V,x}}{\eta}\right)^{\eta}\left[\frac{(1+\tau)\overline{P}_Q}{1-\eta}\right]^{1-\eta}$, and $m_y = P_{V,y} = \frac{1}{A^Y}\left(\frac{rP^K}{\beta}\right)^{\beta}\left(\frac{W}{1-\beta}\right)^{1-\beta}$ are the unit costs in each sector.

It is straightforward to derive the demand functions for primary production factors:

$$K^V = \alpha \frac{P_{V,x}}{rP^K} \cdot V \tag{14}$$

$$K^Y = \beta \frac{P_{V,y}}{rP^K} \cdot Y \tag{15}$$

$$L^V = (1-\alpha)\frac{P_{V,x}}{W} \cdot V \tag{16}$$

$$L^Y = (1-\beta)\frac{P_{V,y}}{W} \cdot Y \tag{17}$$

In sector X, profit maximization requires:

$$m_x = \overline{P}_X \tag{18}$$

In sector Y, each oligopolistic firm i maximizes profits taking the behavior of all other competitors as given:

$$\max_{Y_i} \quad \Pi_i^Y(Y_i) = P_Y(Y) \cdot Y_i - m_y \cdot Y_i \tag{19}$$

The condition for optimality is:

$$\frac{dP_Y}{dY}\frac{dY}{dY_i} \cdot Y_i + P_Y(Y) = m_y \tag{20}$$

Since all oligopolists are equal, condition (20) together with (5) gives the optimal output quantity at the sectoral level:

$$Y = (1-\varphi)\frac{N-1}{N} \cdot \frac{I}{m_y} \tag{21}$$

where $N > 1$ must hold for a positive supply.

3.3 Foreign Trade

Foreign trade includes exports of the homogenous commodity X and imports of intermediate O. Since technology (6) in sector X is Cobb Douglas, imports of O are essential and could not be zero.[12] The economy uses exports of sector X to finance the import of intermediates in the same sector. Tariffs on imports of O are the only form of foreign trade distortion.

3.4 Market Clearing Conditions and Walras Law

There are two factor markets, and two commodity markets in this economy. Equilibrium on factor markets requires

$$K^V + K^Y = \overline{K} \tag{22}$$

and

$$L^V + L^Y = \overline{L} \tag{23}$$

Walras' Law implies balanced trade

$$\overline{P}_X \cdot \left(X^S - X^D\right) = E\overline{P}_O \cdot O \tag{24}$$

where the difference $\left(X^S - X^D\right)$ denotes positive exports by sector X. Moreover, we keep things simple by assuming that the entire production of Y is sold to domestic consumers.

Since (24) is redundant, the SOOE is represented by a system of seven independent equations in eight variables. These are three good quantities, X^S, X^D, Y, the foreign intermediate O, the price of the oligopolistic good P_Y, the factor prices W and rP^K, and the nominal exchange rate E. Two equations describe consumer demand for each good, two equations represent domestic firms' supply, two equations are primary inputs' market clearing conditions, and one equation is the optimal demand for intermediate O. A unique solution is obtained by choosing the nominal exchange rate to be the numéraire, i.e. $E = 1$.

[12]Note that the economy may become autarkic if the technology in sector X is generalised to one with constant elasticity of substitution (CES).

4 Results

Private utility U can be expressed as a function of the tariff rate τ and of the number of oligopolistic firms N. To see this, insert model solutions for consumption demand (29) and (30) in Eq. (1), and obtain the indirect utility function as:

$$U(\tau, N) = \Upsilon \cdot \frac{\left(\frac{N-1}{N}\right)^{1-\varphi} [P_{V,x}(\tau)]^{\varphi} \cdot T(\tau)}{\left[\alpha + \beta \frac{N-1}{N} \cdot T(\tau)\right]^{\varepsilon} \left[(1-\alpha) + (1-\beta)\frac{N-1}{N} \cdot T(\tau)\right]^{1-\varepsilon}} \tag{25}$$

where

$$\Upsilon := \left(\overline{K}\right)^{\varepsilon} \left(\overline{L}\right)^{1-\varepsilon} \left[\alpha^{\alpha}(1-\alpha)^{1-\alpha} \frac{\varphi}{1-\varphi} \frac{A^H}{\overline{P}_X}\right]^{\varphi} \left[\beta^{\beta}(1-\beta)^{1-\beta} A^Y\right]^{1-\varphi}$$

and $\varepsilon := \varphi(\alpha - \beta) + \beta$, and

$$T(\tau) := \frac{1-\varphi}{\varphi}\left(1 + \frac{1-\eta}{\eta}\frac{\tau}{1+\tau}\right). \tag{26}$$

Here, the price of value added in sector X is

$$P_{V,x}(\tau) = \eta \left[A^X \cdot \overline{P}_X \left(\frac{1-\eta}{\overline{P}_O}\right)^{1-\eta}\right]^{\frac{1}{\eta}} (1+\tau)^{-\frac{1-\eta}{\eta}} \tag{27}$$

Note that the condition $\tau > -\eta$, which ensures a positive utility, follows directly from the assumption of positive tariffs.

We will now use (25) to show that deeper globalization may fail to improve national welfare, if the economy is oligopolistic. We will show that there exists an optimal level τ^* of tariffs, below which the economy loses from globalization while the opposite applies if tariffs are higher than that level.

Proposition 1 Optimal level of economic integration in oligopoly
If $N > 1$ is finite, the optimal level of tariffs τ^* is unique and strictly positive.

Proof See appendix.

According to Proposition 1, globalization benefits a country only above a certain threshold of tariffs if the economy is oligopolistic. In stark contrast to standard trade theory, welfare (measured by private utility) is not a monotonously decreasing function of tariffs. Figure 3 reports utility as a function of the tariff rate for the cases $N = 2$ (bold), $N = 4$ (broken), and $N = 8$ dotted.[13]

[13]The calibration used for Figs. 3 and 4 is $\overline{K} = 60$, $\overline{L} = 25$, $A^X = 0.8$, $A^V = A^Y = 1$, $\varphi = 0.2$, $\alpha = 0.33$, $\beta = 0.4$, $\eta = 0.4$, $\overline{P}_X = \overline{P}_O = 1$.

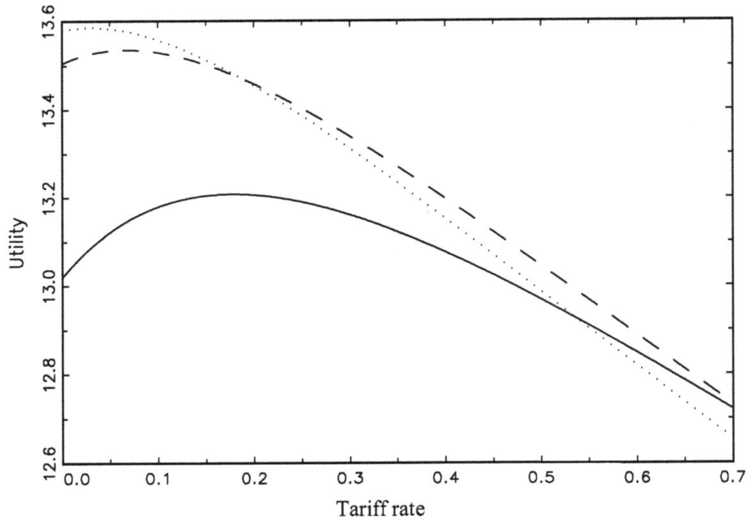

Fig. 3 Private utility as a function of the tariff rate

The intuition for the inverted U-shaped dependence of welfare on the tariff level in the SOOE model is as follows: Resources are limited and production in one sector has opportunity costs in terms of output in the other sector. Due to imperfect competition in the Y-sector, sector Y underproduces and sector X overproduces relative to the efficient (first best) allocation under perfect competition. Since marginal utility is too low for the X good and too high for the Y good a reallocation of resources from the overproducing to the underproducing sector will—other things equal—lead to higher utility.

The same mechanism holds in this case. If tariffs are sufficiently low, imperfect competition in the oligopolistic sector will result in relatively lower production of Y and higher production of X than under a hypothetical scenario with zero tariffs and perfect competition. Hence, if tariffs decrease slightly, imports of the foreign intermediate increase. Due to balanced trade, exports increase as well. This requires more production in sector X with higher demand for domestic resources. The price of primary factors increases. Since the price of good X is exogenously fixed, there is a substitution effect from good Y to good X. This means a welfare decrease. In this setting, the marginal benefit of lower tariffs is more than offset by the marginal damage of a decrease in production of Y.

If however tariffs are high, i.e. higher than the threshold, the balance is distorted in the opposite way, i.e. the ratio of X^D to Y is lower than in the efficient allocation. Thus, the marginal damage of imperfect competition is lower than the marginal damage of high tariffs. In this case, the economy would gain from lower tariffs.

This effect is also visible in Fig. 3. For low levels of the tariff rate, the more firms are active in sector Y, the higher is welfare. However, if tariffs are high, a higher number of firms in this sector may lead to an excessive use of resources in

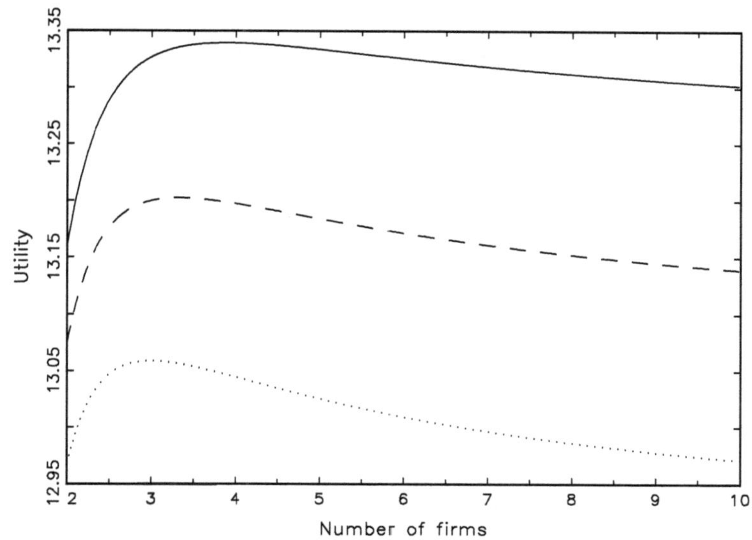

Fig. 4 Private utility as a function of the number of firms

this sector and a decrease in competition would actually increase welfare. Note, for example, that if the tariff rate amount to 60 %, four firms would be welfare-better than eight.

Let us now consider competition policy under the assumption of a given level of tariffs. For simplicity, we will allow N to be any real number greater than one, i.e. we do not require N to be an integer[14]:

Proposition 2 Optimal competition policy under imperfect economic integration

If $\tau > 0$ and finite, ($\tau = 0$) the optimal number of firms is unique and finite (infinite).

Proof See appendix.

According to Proposition 2, perfect competition is not desirable if economic integration is imperfect and tariffs are positive. Welfare as a function of the number of oligopolists does not monotonically increase in the number of firms, as standard theory would suggest. Rather, welfare is inverted U-shaped and there exists an optimal number of oligopolistic firms $0 < N^* < \infty$. Figure 4 reports welfare as a function of the number of firms for the cases $\tau = 0.3$ (bold), $\tau = 0.4$ (broken), and $\tau = 0.5$ (dotted).

The optimal number of oligopolistic firms is

$$N^* = \frac{1}{1-\eta} \cdot \left(1 + \frac{\eta}{\tau}\right) \tag{28}$$

[14]See Beverelli and Mahlstein (2011) for the same assumption.

Clearly, the optimal number of oligopolists is infinite only in the case of perfect economic integration (zero tariffs).

The non-monotonicity of welfare with respect to N is based on the same intuition as in the case of Proposition 1. An increase in the number of firms in the oligopoly means a resource shift towards sector Y. Above the optimal value of N, employed resources and produced output become excessive and an inefficiency arises. However, if globalization improves, the number of firms, which can operate in the oligopolistic sector without efficiency loss becomes higher. Equation (28) provides evidence for the need of a stricter competition policy when firms outsource a greater part of their production because of lower tariffs.

5 Conclusions

This chapter deals with the apparent inability of Italy to gain from globalization. The debate has identified two major determinants for this failure, namely the generalized scarce propensity to innovate and the inadequate specialization pattern of the economy. There is widespread agreement concerning the former that the delayed adoption of ICT by Italian firms prevented them from exploiting the full spectrum of opportunities deriving from globalization. At the same time, the historical competitiveness of the most successful Italian sectors (typically, the low-tech ones) was challenged by the emergence of the large developing countries. Their similar production structure and huge cost advantage progressively pushed the formerly successful Italian sectors onto a declining path, and the whole specialization pattern of the Italian economy became rapidly inadequate.

The delayed adoption of ICT and the rapid obsolescence of the specialization pattern are considered two prominent features of the Italian economy, and indeed the literature on the stagnation of the last two decades includes them among the plausible determinants of the crisis. Since the debate on Italy's inability to gain from globalization focuses on these features to find possible explanations for the crisis, we briefly review the most significant explanations in the second section of this chapter. The analysis reveals that an important weakness of the Italian economy is the scarce degree of competition on the internal markets and especially on service markets. Consequently, we focus on this aspect of the economy to propose an alternative explanation for Italy's failure in gaining from globalization. We specify a very stylized model of a small open oligopolistic economy (SOOE) with outsourcing and show that Italy might effectively have suffered from increased economic integration. The model assumes one oligopolistic and one competitive sector, which outsources part of its production abroad. We use this setting to study the welfare effects of globalization in the form of falling tariffs on intermediates. We show that for a given oligopolistic structure of the economy, globalization may fail to improve welfare, if tariffs are sufficiently low and competition is scarce. We also find that perfect competition is not desirable under positive tariffs, and that an optimal competition policy is necessary. In particular, exogenous advances in economic integration might require more competition in order to be beneficial for the economy.

These results are an application of the well-known Lipsey–Lancaster theory of second best. In general, imperfect competition and tariffs generate underproduction, and a change in either of the two types of distortion induces a resource shift between sectors with direct effects on welfare. If the degree of economic integration is extremely low, there may be underproduction in the protected sector independently of the level of competition. Thus, lower tariffs can reduce underproduction and improve welfare. Conversely, if integration reaches high levels, oligopoly is responsible for underproduction in the non-competitive sector, and advances in integration exacerbate it.

The model proposed in this chapter rests on several standard but crucial assumptions. One of them regards the perfectly functioning labor market, which is assumed to clear autonomously. Clearly, this assumption is in net contrast with the reality of the Italian economy, and needs to be taken into account when drawing conclusions based on model results. Under these limitations, this chapter proposes an alternative explanation for Italy's failure to gain from globalization over the last two decades, which pivots on the level of market competition on internal markets, and claims that the costs due to excessive regulation in some markets may have more than offset the benefits of higher economic integration.

Appendix: Proof of Propositions

The proof of both propositions is based on utility function (25) and on the model solutions:

$$X^D = \frac{\varphi}{1-\varphi} A^V \left(\frac{\alpha}{\Omega}\overline{K}\right)^{\alpha} \left(\frac{1-\alpha}{\Psi}\overline{L}\right)^{1-\alpha} \frac{1}{\overline{P}_X} \cdot P_{V,x}(\tau) \cdot T(\tau) \tag{29}$$

$$Y = \frac{N-1}{N} A^Y \left(\frac{\beta}{\Omega}\overline{K}\right)^{\beta} \left(\frac{1-\beta}{\Psi}\overline{L}\right)^{1-\beta} \cdot T(\tau) \tag{30}$$

where $\Omega(\tau, N) := \alpha + \beta \cdot T(\tau) \cdot (N - 1/N)$, and $\Psi(\tau, N) := (1-\alpha) + (1-\beta) \cdot T(\tau) \cdot \cdot (N - 1/N)$. The conditions $N > 1$ and $\tau \geq 0 > -\eta$ $(\tau > -1 + A^X(1-a)^{1/\eta}(\overline{P}_X/\overline{P}_O))$ guarantee positive solutions in the Cobb–Douglas (CES) case.

Proposition 1 We first show that utility (1) is continuous in τ for $\tau \geq 0$. This is immediately seen from the fact that $\Omega(\tau, N)$ and $\Psi(\tau, N)$ are continuous in τ and strictly positive since $T(\tau) > 0$ for any $\tau \geq 0$. Hence, X^D and Y are also continuous in τ. Differentiating Eq. (1) with respect to the tariff rate yields

$$\frac{\partial U}{\partial \tau} = \left[\varphi \frac{1}{X^D} \frac{\partial X^D}{\partial \tau} + (1-\varphi) \frac{1}{Y} \frac{\partial Y}{\partial \tau} \right] \cdot U. \tag{31}$$

If τ goes to infinity, utility is zero since $\Omega(\tau, N)$ and $\Psi(\tau, N)$ are finite and X^D collapses to zero (see Eq. (27)). For $\tau \geq 0$ $0 < U(\tau, N) < \infty, \forall N > 1$. Thus, $U'_\tau = 0$ if and only if the term in square brackets in Eq. (31) is zero. Its opposite is equivalent to the following cubic equation in the level of tariffs:

$$\tau^3 + a \cdot \tau^2 + b \cdot \tau + c = 0 \tag{32}$$

where

$$a := \frac{1}{d} \left\{ \frac{1}{\eta^2} \frac{N-1}{N} \left[\frac{\beta(1-\beta)(1-\varphi)}{\varphi} \left(3\frac{N-1}{N} - \eta \right) + [\alpha(1-\beta) + (1-\alpha)\beta](1+\eta) \right] + \right.$$
$$\left. + \alpha(1-\alpha) \left(\frac{1}{\eta} \frac{N-1}{N} + 2\frac{\varphi}{(1-\varphi)\eta} - 1 \right) \right\}$$
$$b := \frac{1}{d} \left\{ \left[(1-\alpha)(\alpha(1+\eta)+\beta) - \frac{\beta(1-\beta)(1-\varphi)}{\varphi} \left(1 + \eta - 3\frac{N-1}{N} \right) + \alpha(1-\beta) \right] \right.$$
$$\left. \frac{N-1}{N} \frac{1}{\eta} - \alpha(1-\alpha) \left[2 - \frac{\varphi}{(1-\varphi)\eta} \right] \right\}$$
$$c := -\frac{1}{d} \left\{ \frac{1}{N} \left[\alpha(1-\alpha) + \beta(1-\beta) \frac{1-\varphi}{\varphi} \frac{N-1}{N} \right] \right\}$$

where

$$d := \frac{1}{\eta} \left\{ \frac{1}{\eta} \frac{N-1}{N} [\alpha(1-\beta) + (1-\alpha)\beta] + \frac{\alpha(1-\alpha)\varphi}{1-\varphi} + \frac{\beta(1-\beta)(1-\varphi)}{\varphi} \left(\frac{1}{\eta} \frac{N-1}{N} \right)^2 \right\}$$

Note first that $a > 0$, $d > 0$, and $c < 0$, which ensure two negative and one positive solution. (The sign of b is irrelevant.) Let τ^* be the positive solution. In order to prove that the positive solution is a maximum observe that $U'_\tau(0, N) > 0$ because $c < 0$ and Eq. (32) is the opposite of the term in square brackets in (31). Since $U(\tau, N)$ is continuous, and the other roots of Eq. (32) are negative, it follows that $U'_\tau(0, N) > 0$ in $[0, \tau^*)$. The fact that τ^* is a root of a cubic equation with at least two distinct solutions ensures that $U'_\tau(0, N) < 0$ if $\tau > \tau^*$. Thus, τ^* is a utility maximum. This proves Proposition 1.

Proposition 2 We show first that utility function (1) is continuous in N for $N > 1$. This is immediately from the fact that $\Omega(\tau, N)$ and $\Psi(\tau, N)$ are continuous in N and strictly positive for any $N > 1$ and so are X^D and Y. Differentiating the utility Eq. (25) and setting $U'_N(\tau, N)$ equal to zero yields the following quadratic equation in $M := (N-1)/N$:

$$A \cdot M^2 + B \cdot M + C = 0 \tag{33}$$

with

$$\begin{aligned} A &:= -\beta(1-\beta)\varphi \cdot [T(\tau)]^2 \\ B &:= -[\alpha\varphi(1-\alpha) - \beta(1-\varphi)(1-\beta)] \cdot T(\tau) \\ C &:= \alpha(1-\alpha)(1-\varphi) \end{aligned} \tag{34}$$

Since $A < 0$ and $C > 0$ for all feasible model parameters, $(B^2 - 4AC)$ is strictly positive. This ensures the existence of two real and distinct solutions, which are discordant in sign. Since $N_{1,2} = 1/(1 - M_{1,2})$, the negative solution $M_2 = \left(-B - \sqrt{B^2 - 4AC}\right)/2A$ is unfeasible because $N > 1$ must hold. The positive solution is feasible only if $M_1 = \left(-B + \sqrt{B^2 - 4AC}\right)/2A < 1$, which is equivalent to $(A + B + C) > 0$. Replace A, B, C by their definitions and verify that this is a product of positive terms. Since $A < 0$ and $(B^2 - 4AC) > 0$, $(A \cdot M^2 + B \cdot M + C)$ is positive (negative) for $M < M_1$ ($M > M_1$) which proves that $N_1 = 1/(1 - M_1)$ is a utility maximum. Use definitions (34) and (26) to verify that the optimal N is

$$N^* = \frac{1}{1-\eta} \cdot \left(1 + \frac{\eta}{\tau}\right) \tag{35}$$

Observe that if τ becomes zero, N^* is infinite. This proves Proposition 2.

References

Accetturo A, Bassanetti A, Bugamelli M, Faiella I, Finaldi Russo P, Franco D, Giacomelli S, Omiccioli M (2013) Il sistema industriale italiano tra globalizzazione e crisi, Questioni di Economia e Finanza 193. Banca d'Italia, Rome

Amighini A, Chiarlone S (2004) Rischi dell'integrazione commerciale cinese per il modello di specializzazione internazionale dell'Italia. Liuc Papers 150, Serie Economia e Impresa, 37

Barca F (1997) Compromesso senza riforme nel capitalismo italiano. In: Barca F (ed) Storia del capitalismo italiano dal dopoguerra ad oggi. Donzelli, Roma

Bassanetti A, Iommi M, Jona Lasinio C, Zollino F (2004) La crescita dell'economia italiana tra ritardo tecnologico e rallentamento della produttività, Temi di Discussione 539. Banca d'Italia, Rome

Beverelli C, Mahlstein K (2011) Outsourcing and competition policy. J Ind Comp Trade 11:131–147

Bianco M, Giacomelli S, Rodano G (2012) Concorrenza e regolamentazione in Italia. Questioni di Economia e Finanza 123. Banca d'Italia, Rome

Blanchard O, Landier A (2002) The perverse effects of partial labour market reform: fixed-term contracts in France. Econ J 112(480):214–244

Boeri T, Faini R, Ichino A, Pisauro G, Scarpa C (2005) Oltre il declino. Il Mulino, Bologna

Boeri T, Garibaldi P (2007) Two tier reforms of employment protection: a honeymoon effect? Econ J 117(521):357–385

Brandolini A, Cipollone P (2003) Una nuova economia in Italia. In: Rossi S (ed) La Nuova Economia: i fatti dietro il mito, il Mulino, Bologna

Breda E, Cappariello R (2012) A tale of two bazaar economies: an input-output analysis for Germany and Italy. Economia e politica industrial 39(2):43–69

Chari VV, Kehoe PJ, McGrattan ER (2007) Business cycle accounting. Econometrica 75(3):781–836
Christopoulou R, Vermeulen P (2008) Markups in the Euro area and the US over the period 1981–2004 – a comparison of 50 sectors. Working Paper Series, n. 856. European Central Bank
Ciocca P (2007) Ricchi per sempre? Una storia economica d'Italia (1796–2005). Bollati Boringhieri
Ciocca P (2004) L'economia italiana: un problema di crescita. Rivista italiana degli economisti 9(1)(suppl.):7–28
Ciocca P (2010) La specificità italiana nella crisi in atto. Moneta e Credito 63(249):51–58
CNEL (2007) Liberalizzazioni e privatizzazioni. CNEL, Rome
Crettez B, Fagart MC (2009) Does entry improve welfare? A general equilibrium approach to competition policy. J Econ 98:97–118
D'Ippoliti C, Roncaglia A (2011) L'Italia: una crisi nella crisi. Moneta e Credito 64(255):189–227
Daveri F (2002) The new economy in Europe (1992–2001). Oxf Rev Econ Policy 18:345–362
Daveri F (2004) Why is there a European productivity slowdown? CEPS Working Document 205, Brussels
Daveri F, Jona Lasinio C (2006) Italy's decline: getting the facts right. Paper presented at the conference "Nuovi Temi per la Politica Economica", Giornale degli Economisti and Ente Einaudi in Rome (14 November 2005)
Dew-Becker I, Gordon R (2012) The role of labor-market changes in the slowdown of European productivity growth. Rev Econ Inst 3(2)
Faini R (2003) Fu vero declino? L'Italia degli anni Novanta. Il Mulino 6:1072–1083
Faini R, Haskell J, Barba Navaretti G, Scarpa C, Wey J (2005) Contrasting Europe's decline: do product market reforms help? Paper presented at the conference "Oltre il declino", Fondazione Rodolfo De Benedetti (February 2005)
Faini R, Sapir A (2005) Un modello obsoleto? Crescita e specializzazione dell'economia italiana. In: Boeri T, Faini R, Ichino A, Pisauro G, Scarpa C (eds) Oltre il declino. Il Mulino, Bologna
Fachin S, Gavosto A (2010) Trends of labour productivity in Italy: a study with panel co-integration methods. Int J Manpower 31(7):755–769
Federico G, Wolf N (2012) Italy's comparative advantage: a long-run perspective. CEPR Working Paper 8758
Forni L, Gerali A, Pisani M (2010) Macroeconomic effects of greater competition in the service sector: the case of Italy. Macroecon Dyn 14(05):677–708
Fortis M, Quadrio Curzio A (2003) Alle prese con la concorrenza asiatica. Il Mulino 6:1103–1113
IMF (2015) World Economic Outlook database. IMF, Washington
Jona Lasinio C, Vallanti G (2011) Reforms, labour market functioning and productivity dynamics: a sectoral analysis for Italy. Working Papers Luiss Lab 1193, Dipartimento di Economia e Finanza, LUISS Guido Carli
Lipsey RG, Lancaster K (1956) The general theory of second best. Rev Econ Stud 24:11–32
Lucidi F (2012) Is there a trade-off between labour flexibility and productivity growth? Some evidence from Italian firms. In: Addabbo T, Solinas G (eds) Non-standard employment and quality of work. AIEL series in labour economics, Physica-Verlag HD, pp 261–285
Lucidi F, Kleinknecht A (2010) Little innovation, many jobs: an econometric analysis of the Italian labour productivity crisis. Camb J Econ 34(3):525–546
MEF (2011) Documento di Economia e Finanza 2011. III, Programma Nazionale di Riforma
Nardozzi G (2004) Miracolo e declino. Italia tra concorrenza e protezione. Roma – Bari, Laterza
Nardozzi G (2003) The Italian economic miracle. Rivista di Storia Economica 2:139–180
Neary JP (2003) The road less travelled: oligopoly and competition policy in general equilibrium. In: Arnott R, Greenwald B, Kanbur R (eds) Imperfect economics: essays in honor of Joseph Stiglitz. MIT Press, Cambridge, pp 485–500
OECD (2005) Going for growth. OECD, Paris
OECD (2006) Science and Technology Database
OECD (2012) Main science and technology indicators. OECD, Paris

Onida F (1999) Quali prospettive per il modello di specializzazione internazionale dell'Italia? Economia Italiana 3:573–629

Onida F (2004) Se il piccolo non cresce. Piccole e medie imprese italiane in affanno. Il Mulino, Bologna

Orsi R, Turino F (2014) The last fifteen years of stagnation in Italy: a business cycle accounting perspective. Empir Econ 47:469–494

Pagano P, Schivardi F (2003) Firm size distribution and growth. Scand J Econ 105(2):255–274

Parisi ML, Schiantarelli F, Sembenelli A (2006) Productivity, innovation and R&D: micro evidence for Italy. Eur Econ Rev 50(8):2037–2061

Pilat D, Lee F, Van Ark B (2002) Production and use of TIC: a sectoral perspective on productivity growth in the OECD area. OECD Economic Studies, n. 35

Rossi S (2004) Economia italiana: perché la deriva non si muti in declino. Il Mulino 4:639–650

Sterlacchini A, Venturini F (2014) R&D and productivity in high-tech manufacturing: a comparison between Italy and Spain. Ind Innov 21(5):359–379

Toniolo G (2004) L'Italia verso il declino economico? Ipotesi e congetture in una prospettiva secolare. Rivista italiana degli economisti 9(1)(suppl.):29–45

Trento S (2003) Stagnazione e frammentazione produttiva. Il Mulino 6:1093–1102

van Ark B, O'Mahoney M, Timmer MP (2008) The productivity gap between Europe and the United States: trends and causes. J Econ Perspect 22(1):25–44

Vaciago G (2003) Il declino dell'economia italiana. Il Mulino 6:1084–1092

Venturini F (2004) The determinants of Italian slowdown: what do the data say? EPKE working paper 29, NIESR, London

Zotti J, Lucke B (2014) Welfare-optimal trade and competition policies in small open oligopolistic economies. J Int Trade Econ Dev. http://dx.doi.org/10.1080/09638199.2012.742555

Part II
Economic Growth, Technological Change, and Climate Policy

Status, Wealth Distribution, and Endogenous Economic Growth

Jörn Kleinert and Ronald Wendner

1 Introduction

Inequality has been one of the buzzwords at least since the outbreak of the financial crisis in 2007. It has reminded us that neither income nor wealth are evenly distributed, that their distribution has become increasingly uneven in the recent history of nearly all OECD countries and that this inequality and its increase may matter for economic outcome. The OECD (2014), for instance, reports a robust, sizable and statistically significant effect of inequality on GDP per capita in a study of the OECD countries in recent years. Similarly, Ostry et al. (2014) find that lower net inequality (after taxes and transfers) is robustly correlated with faster and more durable growth, for a given level of redistribution. In addition, redistribution appears generally benign in terms of its impact on growth.

We demonstrate a link between (in)equality and growth in a theoretical framework that includes status considerations in individuals' preferences. Individuals concerned with their status in society invest more in human and physical capital. In our approach, the non-symmetric Power law distribution of household wealth yields households at the lower end of the wealth distribution to work harder to improve its status in society, i.e., to invest more than proportionally. The incentive to do so is particularly pronounced in more equal societies.

We use a power-law wealth distribution, because wealth has been empirically shown to be power-law distributed. Very often a power-law distribution is explained by a stochastic growth process with a reflecting border. We also rely on this explanation and an economy with heterogeneous households (with respect to wealth) that grow randomly according to a known distribution. The power-law

J. Kleinert (✉) • R. Wendner
Department of Economics, University of Graz, Universitätsstraße 15, 8010 Graz, Austria
e-mail: joern.kleinert@uni-graz.at; ronald.wendner@uni-graz.at

B. Bednar-Friedl, J. Kleinert (eds.), *Dynamic Approaches to Global Economic Challenges*, DOI 10.1007/978-3-319-23324-6_7

distribution is robust to the random growth of households. With households caring for status, the distribution affects economic growth. We thereby show that in our framework, more equal economies grow faster.

Putting so much attention to heterogeneity and wealth differences we simplify on the household side. We do not study an overlapping generation (OLG) model as Karl Farmer would surely had suggested, but rely on dynastic households with infinite time horizons. We agree with Karl on the merits of the OLG approach but leave the analysis of wealth heterogeneity in an OLG framework for future work.

2 Status

Social distinction or status is an important motivation of human (and non-human) behavior.[1] This was already shown by Darwin (1871), who emphasized *sexual* selection besides *natural* selection. "*To spread across the population, genes of sexual species not only need to survive in their natural and social environment, but also need to be or appear a more attractive mating partner than their same sex competitors.*" (Truyts 2010, p.137).

Clearly, Darwin was not the first to think about social distinction. Philosophers started to comment on social distinction more than 2400 years ago. In his *The Republic* (Book II), Plato argues: "*Since . . . appearance tyrannizes over truth and is lord of happiness, to appearance I must devote myself*". This passage astoundingly resembles Darwin's argument on sexual selection.

More recently, political philosophers and economists regularly discuss social distinction. For example, Smith (1759) in his *Theory of Moral Sentiments* wrote: "*The poor man's son. . . when he begins to look around him, admires the condition of the rich. He finds the cottage of his father too small . . . It appears in his fancy like the life of some superior rank of beings, and, in order to arrive at it, he devotes himself for ever to the pursuit of wealth and greatness.*" (Smith 1759, p. 181) Likewise, Veblen argues "*Conspicuous consumption of valuable goods is a means of reputability to the gentleman of leisure.*" (Veblen 1899, p. 64).

But what is social distinction or status? Consulting the modern literature, many different answers are found. Psychologists and behavioral economists have established that individuals experience happiness by doing well relative to some reference group. This motivation for behavior might be called the *distinction effect*. However, households not only care about distinction, but they also care about inequality. As has been shown, individuals dislike being "too different" from their

[1]Different authors employ various terms, with slightly varying meanings, to describe positional preferences. These terms include *(negative) consumption externality, relative wealth or consumption, jealousy, envy, keeping or catching up with the Joneses, external habits, positional concerns, conspicuous wealth or consumption*. In this article, we use these terms synonymously, though we focus on *relative wealth*.

peers. For example, Fehr and Schmidt (1999), find that people dislike income inequality, but they are more upset when their own income falls short than they are pleased by an excess, compared to their reference level. Cowan et al. (2004) argue that some activities become more desirable when they can be shared with a group of peers (peer group effect). Other activities become more desirable if they allow the consumer to emulate the consumption of an elite group that he or she aspires to join (aspiration effect). Still other activities become more desirable when the individuals can, through wealth or personal endowments, outshine their peers (distinction effect). In this article, we adopt the following view:

- Households care about their rank in the wealth distribution (Mujcic and Frijters 2013).

This view primarily captures the distinction effect. In addition, as the normalized wealth distribution is not symmetric (as argued below), this view also captures to some extent the reasoning by Fehr and Schmidt (1999). The disutility incurred from its position in the wealth ranking decreases more than proportionally with the wealth of a household.

There is abundant empirical evidence that corroborates the desire for social distinction or status. The significant studies of biologists and psychologists aside, there are three types of study in economics that convincingly demonstrate the significance of *social distinction or status*.

The first type of study is related to happiness studies (based on census data). This type of study was pioneered by Easterlin (1995) who observed that a significant rise of income over the last 50 (or so) years was not accompanied by a corresponding rise in happiness. This finding of a lack of trend in happiness is robust over time and over countries. One factor explaining this lack of a trend in happiness is the fact that *relative* income (or wealth) was stationary over the observation period. If households derive happiness not from the absolute, but from the relative level of income, then a rise in average income need not be accompanied by a rise in happiness. However, this finding does not hold in a cross-sectional context. At a given point in time, those households with higher wealth report a higher level of happiness than households with lower wealth (Oswald 2004). These results were confirmed by many other studies (cf. Alvarez-Cuadrado et al. 2012; Ferrer-I-Carbonell 2005; Luttmer 2005).

The second type of studies are stated preference studies. Such studies are typically designed as lab experiments. Respondents face hypothetical situations and need to state their preferred choices (Alpizar et al. 2005; Johansson-Stenman et al. 2002; Johansson-Stenman and Martinsson 2006; Solnik and Hemenway 1998, 2005). Typical questions posed in such studies are the following.

Suppose you can choose between two situations A or B. Which one do you prefer:

A: Everything else equal, your yearly income is Euro 100,000, while everybody else's income equals Euro 150,000;

B: Everything else equal, your yearly income is Euro 80,000, while everybody else's income equals Euro 50,000.

Consistently, about a third of respondents prefer situation B over situation A, in spite of the lower absolute income—but in face of the higher relative income.

The third type of study involves revealed preference. Of course, natural experiments are rare in this context. However, some studies show that social distinction is decisive for many economic decisions. For example, Stark and Taylor (1991) studies migration and finds that if absolute income is controlled for, relative income (within the village) is important in explaining migration decisions (Truyts 2010, p. 140).

In the recent past, these insights from behavioral economics were employed in a variety of contexts in economics: happiness, (optimal) tax policy, economic growth, optimal redistributive taxation, asset pricing, public good provision, or natural resource extraction, just to name a few. In this article, we aim at analyzing the impact of wealth distribution—when households have a preference for social distinction—on (fully) endogenous growth.

This research question is not only motivated by the recent discussion about growth and inequality, as forcefully delineated by Piketty (2014). It is also motivated by a standing discussion about the theme. Alesina and Rodrik (1994), Persson and Tabellini (1994), Bertola (1993), present prominent examples. What distinguishes our article from theirs, however, is the following fact:

- We harness a stochastic framework in which the relative wealth (consumption) distribution essentially follows a power-law (cf. Gabaix 2009; Reed 2003).

In the following sections, we adopt the view that households care about their ranks in the wealth distribution. We demonstrate that the relative wealth distribution necessarily follows a power law, independently of the specification of social distinction. In the following section, we analyze the impact of this wealth distribution on endogenous growth.

3 The Distribution of Relative Wealth

Wealth is not evenly or normally distributed. Its distribution is better described by a power-law. Power-laws are characterized by fat tails. With respect to wealth they comprise also the extremely wealthy part of the population. Pareto (1897) noted that 80 % of the land in Italy was owned by 20 % of the population. He found this ratio in more empirical distributions. With respect to income, he proposed that the probability that income exceeds a particular value x is well described by $1/x^k$, where k is approximately 1.5. Thus, he described income and wealth by power-law distributions. That has proven successful since then in numerous studies.

Power law distributions arise as a consequence of a random growth process with a reflecting lower barrier, i.e., minimum wealth w_{min} in our framework (see Gabaix 2009 or Jones 2014 for a discussion). We depart from the existing literature on wealth distribution and growth only in that we apply a stochastic framework, where future returns on capital are uncertain at the point of investment. The households only know the distribution of return r.

There is a unit mass of (consumer-producer) households. Production technology is given by a Cobb-Douglas production function according to which a homogeneous output is produced by individual capital along with aggregate capital (giving rise to a positive production externality). Labor is inelastically supplied and normalized to unity.

Aggregate capital is a means to capture technological progress. In particular, aggregate capital captures *disembodied learning from net investment* (cf. Groth and Wendner 2014). Here, we specify aggregate capital in a way such that the production externality gives rise to *full* (rather than semi-) endogenous growth.

$$y_{h,t} = a_{h,t} K_t^{1-\alpha} k_{h,t}^{\alpha} \tag{1}$$

Production $y_{h,t}$ is specific to household h and to time t.

Yet, while households have different and time-variant outputs, the output of the "average" household describes the macro-behavior of the model almost completely. The average household's production function is given by

$$y_t = a_t K_t^{1-\alpha} k_t^{\alpha} \,. \tag{2}$$

This production function is commonly used in AK-models of economic growth. One specific example that is related to our model is discussed by Corneo and Jeanne (2001). In (2)

$$a_t \equiv \int_{h=0}^{1} a_{h,t}\, dh\,, \quad k_t \equiv \int_{h=0}^{1} k_{h,t}\, dh\,.$$

A household qualifies in our model as an average household at time t if its wealth $w_{h,t}$ describes the first moment of the wealth distribution in t. Therefore, at different dates t, *different* households qualify as the *average household*.

In order to study the wealth distribution, for each date t, we normalize the wealth of each household h by the average wealth $w_t \equiv \int_{h=0}^{1} w_{h,t}\, dh$. A household's normalized wealth is given by

$$\omega_{h,t} = w_{h,t}/w_t \,,$$

where $\omega_{h,t}$ describes *relative wealth*, and the normalized wealth of the average household is equal to one. Normalization is required to ensure stationary of the relative wealth distribution in our framework with endogenous growth (increasing absolute wealth).

With stochastic returns to capital in each period, relative wealth of the household is likely to change even if the distribution is stationary. On average, large wealth grows at the same rate as small wealth. Yet, there is a lower bound of wealth ensuring that wealth cannot fall below a particular threshold. Debt for instance cannot grow infinitely since no lender would be willing to finance additional needs. The stochastic process describing the evolution of wealth is best expressed as a Gaussian diffusion process

$$d\omega_{h,t} = \mu(\omega_{h,t})dt + \sigma(\omega_{h,t})dz_t\,, \tag{3}$$

where μ, the increments of wealth with $\mu(\omega_{h,t}) = g_r\omega_{h,t}$, is the drift parameter, σ, the standard deviation of the increments with $\sigma(\omega_{h,t}) = v\omega_{h,t}$, is the diffusion parameter and z_t denotes a Brownian motion. Parameters g_r and v characterize the stochastic process of the evolution of *relative* wealth above the minimum wealth level. All changes $d\omega_{h,t}$ that would push the household below the minimum level do not materialize and bind household wealth at the minimum level.[2]

Let $f(\omega, t)$ denote the distribution of relative wealth in t. The drift parameter of the process changes because the changes in wealth are measured relative to the changes in average wealth. The rate of relative change g_r is not necessarily zero as the reflecting lower barrier, the minimum wealth $w_{min,t}$ biases low wealth levels upwards. The average growth of unconstraint relative wealth is therefore negative. Describing the evolution of wealth in this way allows to harness the Kolmogorov forward equation to determine the future distribution of relative wealth $f(\omega, t)$.

$$\frac{\partial}{\partial t}f(\omega,t) = -\frac{\partial}{\partial \omega}\left[\mu(\omega,t)f(\omega,t)\right] + \frac{\partial^2}{\partial \omega^2}\left[\frac{\sigma^2(\omega,t)}{2}f(\omega,t)\right] \tag{4}$$

In equilibrium the distribution must be stationary which requires $\frac{\partial}{\partial t}f(\omega,t) = 0$ or

$$\frac{\partial}{\partial \omega}\left[\mu(\omega,t)f(\omega,t)\right] = \frac{\partial^2}{\partial \omega^2}\left[\frac{\sigma^2(\omega,t)}{2}f(\omega,t)\right] \tag{5}$$

To continue, we reason—as argued above—that wealth follows a Pareto distribution $f(\omega) = b\omega^{-\kappa-1}$ where $b = \omega_{min}^{\kappa}\,\kappa > 0$, and $\kappa > 1$ denote constants.[3]

Inserting this assumption into (5) yields

$$\frac{\partial}{\partial \omega}\left[g_r\omega b\omega^{-\kappa-1}\right] = \frac{\partial^2}{\partial \omega^2}\left[\frac{v^2\omega^2}{2}b\omega^{-\kappa-1}\right],$$

$$-\kappa g_r b\omega^{-\kappa-1} = (-\kappa+1)(-\kappa)\left[\frac{v^2 b}{2}\omega^{-\kappa-1}\right],$$

$$g_r = (1-\kappa)\frac{v^2}{2}\,, \tag{6}$$

[2]We may consider a government that redistributes income—in a budget neutral way—so as to ensure a minimum income of the poorest households.

[3]Observe that the cumulated distribution function is given by $F(\omega) = 1 - (\omega/\omega_{min})^{-\kappa}$.

which allows us to infer the shape parameter of the distribution:

$$\kappa = 1 - 2g_r/v^2. \tag{7}$$

Equation (7) endogenizes the wealth distribution in that the shape parameter is determined by the drift and the diffusion parameter of the random growth process. Thereby, the growth rate of relative wealth, g_r, reduces κ which makes the wealth distribution more unequal. A higher diffusion parameter, in contrast, corresponds c.p. to a higher shape parameter, i.e., less inequality. Note, that the growth rate of relative wealth is negative, since $1 - \kappa$ is negative. The reflecting barrier "corrects" some particularly negative growth rates and thereby raises the average growth rate with barrier above the average growth rate without barrier. Hence, the level of the growth rate of relative wealth displays the importance of the reflecting barrier—the economy's minimum wealth $w_{min,t}$. The higher this minimum threshold, the lower is the growth rate of relative wealth and the more "equal" c.p. is the wealth distribution.

4 Relative Wealth Distribution and Endogenous Growth

After having established stationarity of the relative wealth distribution, cf. Eqs. (4)–(7), we examine the distribution's impact on economic growth. In what follows, we closely follow Corneo and Jeanne (2001). However, we assume a stochastic process describing the evolution of the economy: future values of the variables of interests are necessarily unknown. The economy is populated by a unit mass of infinitely-lived households, indexed by h, with identical preferences represented by the following intertemporal utility function:

$$\mathbf{E}_0[U_h] = \int_0^{\infty} E_t\left[\ln(c_{h,t}) - s\ln(\rho_{h,t})\right] e^{-\beta t} dt\,, \quad \beta > 0\,, \tag{8}$$

where $c_{h,t}$ denotes the consumption of household h at time t, s is the importance of the status in the utility of a household, $\rho_{h,t} \in (0, 1]$ denotes the distance of the household to the wealthiest household of the economy, and β is the pure rate of time preference.

The term $[s\ln(\rho_{h,t})]$, represents the utility derived by a household as a consequence of its wealth difference to the poorest household, as given by the rank in the wealth distribution. This rank is given by the value of the cumulated distribution function $F(\omega_{h,t}) = 1 - (\omega_{h,t}/\omega_{min,t})^{-\kappa}$. We define $\rho_{h,t}$ as

$$\rho_{h,t} = \left(\frac{w_{h,t}}{w_{min,t}}\right)^{-\kappa}, \quad 0 < w_{min,t} \leq w_{h,t}\,. \tag{9}$$

In (9), a household's rank $\rho_{h,t} \in (0, 1]$ is just a weighted ratio of that household's wealth to the minimum wealth w_{min}. The rank of a household, i.e., the distance to the wealthiest household strictly decreases in wealth.

A household uses its wealth to finance consumption and investment, the latter of which affects its future wealth positions. The maximization of the utility function (8) is subject to the stochastic period wealth accumulation constraint

$$dw_{h,t} = r_t w_{h,t} dt + \sigma w_{h,t} dz - c_{h,t} dt \tag{10}$$

where r_t is the average real return of wealth at time t, and σ^2 is the instantaneous rate of variance per unit of time. If $\sigma = 0$ the deterministic case results. In any case, as

$$k_{h,t} = w_{h,t} \quad k_t = w_t\,, \quad K_t = k_t\,, \tag{11}$$

it is the change in wealth of the average household, dw_t, that determines growth in the economy, cf. Eqs. (1) and (2).

The expected change in wealth and capital that follows the process in Eq. (10) is given by $\mathbf{E}_t dk_{h,t} = (r_t k_{h,t} - c_{h,t})dt$. For the following derivations, we also state $\mathbf{E}_t dk^2_{h,t} = \sigma^2 k^2_{h,t} dt$.

To maximize utility (8), a household chooses optimal consumption level $c^*_{h,t}$ so as to maximize the continuous-time stochastic Bellman equation. Consumption $c^*_{h,t}$ is necessarily a random variable. Given our setup in which the random element enters through the simple diffusion process that describes the change in wealth, only the constraint includes stochastic elements. Ignoring subindexes for the moment, the solution of the maximization problem is found where the risk-adjusted current value Hamiltonian is maximized (at all t), and the costate variable λ obeys the stochastic differential equation

$$d\lambda = \lambda\beta dt - \frac{\partial H}{\partial k}(c^*, \lambda, k, \xi)dt + \xi dz\,. \tag{12}$$

Costate variable λ reflects the shadow price of capital in terms of expected utility. $\xi_{h,t}$ denotes the instantaneous conditional covariance of $\lambda_{h,t}$ and z_t, i.e $\xi_{h,t} = (E_t d\lambda_{h,t} dz)\,/dt$.

The risk adjusted current value Hamiltonian is given by

$$H_h(c_{h,t}, k_{h,t}, \lambda_{h,t}, \xi_{h,t}) = U(c_{h,t}, \rho_{h,t}, k_{h,t})dt + \lambda_{h,t}[r_h k_{h,t} - c] + \xi_{h,t}\sigma k_{h,t}$$

The Hamiltonian is maximized for all time given the costate variable λ_t, the covariance ξ_t and the capital stock k_t if the consumption level c^* satisfies

$$\frac{\partial H_h}{\partial c_h}(c^*_h, \rho_h, k_h, \lambda_h, \xi_h) = U_c(c^*_h, k_h) - \lambda_h = 0 \tag{13}$$

Given our utility function (8), we have $\lambda_h = 1/c_h$. Applying Ito's lemma to the costate variable yields

$$d\lambda_h = -\frac{1}{c_h^2}dc_h + \frac{1}{c_h^3}dc_h^2 \tag{14}$$

However, inserting the first derivative of the Hamiltonian with respect to capital/wealth in the stochastic differential equation for the costate variable, we also get an expression for the change of the costate variable on the balanced growth path as it is given in Eq. (12)

$$d\lambda_h = \lambda_h \beta dt - \left[\frac{\partial U_h(c_h^*, k_h)}{\partial k_h} - \lambda_h r + \xi_h \sigma\right] dt + \xi_h dz$$

In expectation that yields

$$\frac{\mathbf{E}_t d\lambda_h}{dt} = \lambda_h \left[\beta - r + R(k_h)\sigma^2\right] - \frac{s\kappa}{k_h} \tag{15}$$

where $R(k)$ denotes the relative risk aversion $R(k) = \frac{\xi_h}{\sigma \lambda_h}$.

To solve the problem we make the guess that consumption on the balanced growth path is a fixed share of wealth/capital, i.e., $c_h(k) = \eta k_h = \eta w_h$ and verify this solution. With $c_h(w_h) = \eta w_h$ we immediately have $\mathbf{E}_t dk = (r - \eta)kdt$ and $\mathbf{E}_t dc = (r - \eta)cdt$. Also, $\mathbf{E}_t dc^2 = \sigma^2 c^2 dt$. Moreover, the change in the costate variable $d\lambda_h$ as given in (14) reads after "dividing by" dt

$$\frac{\mathbf{E}_t d\lambda_h}{dt} = -\frac{1}{c_h}[r - \eta] + \frac{1}{c_h}\sigma^2 = \lambda_h[r - \eta] + \lambda_h \sigma^2 \tag{16}$$

From Eqs. (15) and (16) follows $\eta - r + \sigma^2 = \left[\beta - r + R(k)\sigma^2\right] - s\eta\kappa$. Solving for η yields

$$\eta = \frac{\beta + (R(k) - 1)\sigma^2}{1 + s\kappa} \tag{17}$$

Note that η is not household-specific. All right-hand side variables are the same for all households. With a relative risk of one as in our case as we show below, η equals $\eta = \frac{\beta}{1+s\kappa}$. If additionally relational preferences do not play a role (s=0), η equals the time preference rate β.

To verify that optimal consumption is a fixed share of capital/wealth $c_h = \eta k_h$ we use the value function $J(k_0)$. It denotes the maximized value of the utility function (8) at the time when planning starts in $t = 0$. Using the evolution of wealth/capital (10), we express the capital stock k_t in terms of the capital stock k_0 from the beginning of the planning period, i.e., $k_t = k_0 e^{(r-\eta+\sigma^2/2)t+\sigma(z_t - z_0)}$. The value function J is expressed in state variable k. In our case $J(k)$ reads

$$\begin{aligned}
J(k_0) &= \mathbf{E}_0\left\{\int_0^\infty e^{-\beta t}\ln \eta k_{h,t} - s\ln\left(\left[\frac{k_{min,t}}{k_{h,t}}\right]^\kappa\right)dt\right\} \\
&= \mathbf{E}_0\left\{\int_0^\infty e^{-\beta t}\ln \eta k_{h,0} - s\ln\left(\left[\frac{k_{min,0}}{k_{h,0}}\right]^\kappa\right)e^{(r-\eta+\sigma^2/2)t+\sigma(z_t-z_0)}dt\right\} \\
&= \ln \eta k_{h,0} - s\ln\left(\left[\frac{k_{min,0}}{k_{h,0}}\right]^\kappa\right)\int_0^\infty e^{-\beta t}e^{(r-\eta+\sigma^2/2)t}dt \\
&= \frac{\ln \eta k_{h,0} - s\ln\left(\left[\frac{k_{min,0}}{k_{h,0}}\right]^\kappa\right)}{\beta - r + \eta - \sigma^2/2}
\end{aligned} \tag{18}$$

Taking the first and the second derivative with respect to the state variable k, we can verify that the rate of relative risk $R(k_h) \equiv -\frac{J''(k_h)k_h}{J'(k_h)}$ is one. That of course results from the log structure of the utility function. Finally, the value function $J(k)$ is a linear function of $\ln k$.

Moreover, on the long-run equilibrium growth path, the expected interest rate is constant. On the balanced growth path, it equals the expected growth rate of wealth and consumption. The steady state growth rate of consumption μ is given by

$$\mathbf{E}\mu = r - \eta = r - \frac{\beta}{1+s\kappa} \tag{19}$$

The growth rate μ is positively affected by the expected interest rate E[r], i.e., by the marginal product of capital (wealth) in the economy, which depends on the productivity parameter a. The growth rate depends positively on κ, $\frac{\partial\mu}{\partial\kappa} = \frac{s\beta}{(1+s\kappa)^2} > 0$. Thus, the shape parameter of the cumulative distribution function κ is also positively related to the growth rate.[4] The reason is that the distance to the maximum wealth is the lower the higher the shape parameter is. Utility from distance to the poorest household is the larger, the higher the shape parameter is, i.e., $\frac{\partial U_\rho}{\partial w_{h,t}} = \frac{s\kappa}{w_{h,t}}$. Households do therefore invest more, the more equal society is which results in a higher growth rate μ.

Finally, we have a look at the role of the positional preferences in the model. Without this preference structure (if $s = 0$), the model is an endogenous growth model with growth rate $\mu = E[r] - \beta$. Considerations concerning the income distribution do not matter. If they matter (i.e., if $s > 0$), the growth rate increases in the strength of the positional part in the preferences, i.e., $\frac{\partial\mu}{\partial s} = \frac{\beta}{(1+s\kappa)^2} > 0$.

[4] The relationship between the shape of the distribution function and the Gini-coefficient is given in Appendix.

5 Discussion of the Results

Growth and inequality are non-ambiguously positively related in this model. There is empirical evidence that this has been true in the OECD countries in the comprehensive overview given by OECD (2011), whereby particularly skill-biased technical change has been blamed for the positive relationship between growth and inequality. In our approach, we do not have labor at all, hence skill-biased technical change cannot be the reason. Instead, the positive relationship between growth and inequality results from average growth in our random wealth-growth process. The counter force is the diffusion parameter which characterizes how likely it is that winners and losers in a particular period switch positions in the next period. Such movements in relative wealth over time are best reflected in measures of the vertical income distribution of households.

With respect to vertical income distribution, the latest trends are alarming, also in Austria and Germany. The schooling system selects early and often "household education persisting" weakens the great driver of reduced inequality in the second half of the twentieth Century. During these years, education of low-income households had increased a lot, allowing for higher incomes of the younger generations relative to the older generation. This process seems to have slowed or even stopped. In our model that goes along with a lower diffusion parameter and more inequality in the long-run.

Appendix

The Gini coefficient Γ of a continuous distribution $F(w)$ is given by

$$\Gamma(w) = 2\int_{w_{min}}^{\infty} F(w)(1 - F(w))dw\frac{1}{2\mu},$$

where μ denotes the distribution's mean. If the distribution $F(w)$ follows a power-law with $F(w) = \left(\frac{w_{min}}{w}\right)^{\kappa}$ and $w_{min} \leq w$, $\mu = \frac{\kappa}{\kappa-1}w_{min}$ and the equation above is solved by

$$\begin{aligned}\Gamma(w) &= \int_{w_{min}}^{\infty} \left(\frac{w_{min}}{w}\right)^{\kappa} - \left(\frac{b}{w}\right)^{2\kappa} dw\frac{1}{\mu} \\ &= \left[-\frac{1}{1-\kappa}w_{min} + \frac{1}{1-2\kappa}w_{min}\right]/\mu \\ &= \frac{\kappa}{(1-\kappa)(1-2\kappa)}w_{min}\frac{\kappa-1}{\kappa w_{min}},\end{aligned}$$

yielding the Gini coefficient

$$\Gamma(w) = \frac{1}{2\kappa - 1}.$$

References

Alesina A, Rodrik D (1994) Distributive politics and economic growth. Q J Econ 109:465–490

Alpizar F, Carlsson F, Johansson-Stenman O (2005) How much do we care about absolute versus relative income and consumption? J Econ Behav Organ 56:405–421

Alvarez-Cuadrado F, Casado JM, Labeaga JM, Sutthiphisal D (2012) Envy and habits: panel data estimates of interdependent preferences. Banco de Espana working paper 1213. Available at SSRN: http://ssrn.com/abstract=2014005+

Bertola G (1993) Factor shares and savings in endogenous growth. Am Econ Rev 83:1184–1198

Cowan R, Cowan W, Swann GMP (2004) Waves in consumption with interdependence among consumers. Can J Econ 37:149–177

Corneo G, Jeanne O (2001) Status, the distribution of wealth, and growth. Scand J Econ 103(2):283–293

Darwin C (1871) The descent of men and selection in relation to sex. John Murray, London

Easterlin RA (1995) Will raising the incomes of all increase the happiness of all? J Econ Behav Organ 27:35–47

Fehr E, Schmidt K (1999) A theory of fairness, competition, and cooperation. Q J Econ 114:817–868

Ferrer-I-Carbonell A (2005) Income and well-being: an empirical analysis of the comparison income effect. J Public Econ 89:997–1019

Gabaix X (2009) Power laws in economics and finance. Annu Rev Econ 1:255–293

Groth C, Wendner R (2014) Embodied learning by investing and speed of convergence. J Macroecon 40:245–269

Johansson-Stenman O, Carlsson F, Daruvala D (2002) Measuring future grandparents' preferences for equality and relative standing. Econ J 112:362–383

Johansson-Stenman O, Martinsson P (2006) Honestly, why are you driving a BMW? J Econ Behav Organ 60:129–146

Jones CI (2014) Pareto and Piketty: the macroeconomics of top income and wealth inequality. NBER working paper 20742

Luttmer EFP (2005) Neighbors as negatives: relative earnings and well-being. Q J Econ 120:963–1002

Mujcic R, Frijters P (2013) Economic choices and status: measuring preferences for income rank. Oxf Econ Pap 65:47–73

OECD (2011), Divided We Stand: Why Inequality Keeps Rising. http://www.oecd.org/els/soc/dividedwestandwhyinequalitykeepsrising.htm

OECD (2014) Growth and inequality: a close relationship? http://www.oecd.org/economy/growth-and-inequality-close-relationship.htm

Ostry JD, Andrew B, Tsangarides CG (2014) Redistribution, inequality, and growth. IMF staff discussion note SDN 14/02

Oswald AJ (2004) Well-being over time in Britain and the USA. J Public Econ 88:1359–1386

Pareto V (1897) Cours d'économie politique, Lausanne: F. Rouge

Persson T, Tabellini G (1994) Is Inequality Harmful for Growth? Am Econ Rev 84:600–621

Piketty T (2014) Capital in the twenty-first century. Harvard University Press, Cambridge, MA

Reed WJ (2003) The Pareto law of incomes - an explanation and an extension. Physica A 319:469–486

Smith A (1759) The theory of moral sentiments. In: Raphael DD, Macfie AL (eds) (1976). The theory of moral sentiments. Oxford University Press, Oxford
Solnik SJ, Hemenway D (1998) Is more always better? A survey on positional concerns. J Econ Behav Organ 37:373–383
Solnik SJ, Hemenway D (2005) Are positional concerns stronger in some domains than in others? Am Econ Rev 95:147–151
Stark O, Taylor JE (1991) Migration incentives, migration types: the role of relative deprivation. Econ J 101:1163–1178
Truyts T (2010) Social status in economic theory. J Econ Surv 24:137–169
Veblen T (1899) The theory of the leisure class, in the theory of the leisure class. Houghton-Mifflin edition (1973). Houghton Mifflin, Boston

Technological Change in Information and Communication: Consequences for Science

Wolf Rauch

1 Introduction

Being an Information Scientist I will not and cannot contribute to the economic analysis of Technological Change. However, Technological Change is not only an object of scientific research; it sometimes changes the way scientific research is performed. This may happen if Technological Change creates new tools, which are then used by scientists (like the microscope, the printing press, the computer). In this case Technological Change opens new doors, enables new insights and modifies the way research is done. This may lead to new concepts, new paradigms, and new results.

An even more important impact on the way science is performed occurs when Technological Change modifies the basic structures of information and communication in society. Then the fundamentals of scientific working and scientific thinking may be changed, too.

Information and communication is more than a "veil" which lies over reality. It does not report without bias. Information and communication technology influences how we think about the world. By changing the way we see reality, we change reality. Therefore Technological Change in the information and communication systems offers new material to be studied by science (with our traditional instruments), and it gives us new tools, requires a new organization of work and leads to a new perception of reality.

Much has been said already about the mutual dependence of media and content. Important ideas came from Socrates (Plato 1997), McLuhan (1964), and Habermas (1981), to mention just a few. However, the rise of the internet and the

W. Rauch (✉)
Department of Information Science and Information Systems, Karl-Franzens University Graz, Universitätsstrasse 15, A-8010 Graz, Austria
e-mail: wolf.rauch@uni-graz.at

B. Bednar-Friedl, J. Kleinert (eds.), *Dynamic Approaches to Global Economic Challenges*, DOI 10.1007/978-3-319-23324-6_8

dramatic increase of communicated information ("Big Data") (Mayer-Schönberger and Cukier 2013) lead to a new situation. In my contribution to this book, I want to address three aspects which are currently under discussion in Information Science and will be of importance to all other scientific disciplines as well; namely consequences of Big Data for the social system of science (Sect. 2), the unsolved problem of long term data storage (Sect. 3), and the consequences for scientific thinking itself (Sect. 4). However, this development is not only important for science; it will affect all parts of our society.

2 Changes in the System of Science

These days, the influence of new information and communication technology on science is discussed mainly on its surface: How scientific research is seen and evaluated from outside. Performance records, rankings, impact points, accounting of intellectual output, and other quantitative measures of scientific results gain importance and allow more transparency in scientific work, more visibility to everyone. But this is often done in a rough statistical form. We get an oversimplified picture of scientific disciplines, developments, institutions, or even single scientists. Surprisingly, this picture is in the majority of cases not too wrong. It may be criticised in detail, but it is mostly correct in its main message.

However, there is always interaction between the observer and the object observed; i.e. the system reacts to the way it is being evaluated. For example: As citations and impact points become more valued in scientific work, they enhance publication in established disciplines with high ranked journals. This development hinders multidisciplinary research, new paradigms and structural changes in science. Measurement and quantification need categories and a well-established terminology. They focus necessarily on the past, not on the presence or future. In other words, the new fashion of accountability in science hinders flexibility and innovation. It gets increasingly risky for an academic career to work in new fields. Changes in the fundamental paradigms are therefore less likely to occur.

Another new development regards distribution of scientific work. The process of manipulation of printed books, papers and copies used to be slow and expensive. Now we have a quick and cheap exchange of digital files. Speed and easiness create new qualities in scientific cooperation; they facilitate international team work.

However, new problems we face are intellectual property, authentication, and plagiarism. The traditional system of science is very much based on paper for communication and documentation. Printed publications are the main input, output, and memory of the system of science. If and only if scientific results are published on paper in a generally accepted source, so far, they play a role in science. Only then, they will legitimately be considered for further research or application. Publishing on paper is an important prerequisite to be accepted as part of the scientific community. A scientific book or journal is the main interface to scientific colleagues and to the outside world. The printed publication in an established format may have even legal consequences. In case of an accident (a bridge collapses, a patient dies, a company goes bankrupt), the question of competent behaviour, of responsibility and of guilt is decided on the basis of published texts. We trust in the world of written knowledge since the publication process is highly elaborate and has lots of gatekeepers: Editors, peer reviewers, publishers, lectors, and of course the readers shall guarantee the quality of written texts. The paper based system of universities and other scientific institutions produces the necessary hierarchy of qualification and reputation to organize, control, and reproduce this system.

All these instances of quality assurance and legitimacy are not developed yet in an electronic environment. Everybody can publish ideas at very low cost in the internet; this content will be found and reproduced by information retrieval systems. In the internet there are hundreds of contributions on Flat Earth Theory, on Creationism or Esotericism. The user can never be sure whether an article has been written by the author named, if this person exists at all or what background or qualification he, she or it has. A text could even have been written by a computer program. Contributions may change overnight or disappear at all. A little example happened during the time this article has been written: In the World Soccer Championship of 2014 the Twitter account "@FifNdhs" published forecasts with all possible outcomes of the final game. When the tournament had been finished the owner of the account deleted all but the right version, and pretended to "prove" that the event had been manipulated (DiePresse.com 2014). The fraud has luckily been discovered.

Manipulation happens with texts, but also with pictures. This is even more dangerous since we believe in pictures much more than we trust in written texts. Manipulation blurred the boundary between reality and fiction, between real and virtual world.

Developments like the ones pointed out can do harm to science and its reputation in society. Electronic publication is a far less solid fundament for science than is print; this is true at least up to now. In the new digital information and communication environment the traditional organizational infrastructure of knowledge is in danger, but also its fundament of trust and legitimacy. In the ancient culture of the spoken word, the commonly accepted knowledge was based on a collective memory of basic experiences and skills shared by everyone. It was documented in the form of proverbs, of myths, and legends. In the present culture of the written word, a complex social system of academic institutions, degrees, tests, peer reviews etc. decides what kind of knowledge in society is "lege artis" and what is regarded as pseudo-science. In the information society, these traditional instances of legitimacy vanish. In the internet it is not so easy to distinguish sound science from voodoo. In the long run the confidence in science may fade away in spite of it's so far role in society as the last commonly accepted orientation. Religion, art, politics and tradition have long since lost their credibility.

This development could finally undermine the social coherence due to the Matthew effect. Robert Merton coined the phrase of the Matthew Effect in 1968 (Merton 1968). He derives the name from a verse in the Gospel of Matthew, pertaining to Jesus' parable of the talents: "For to everyone who has will more be given, and he will have an abundance. But from the one who has not, even what he has will be taken away" (Matthew 25:29 2011). This phenomenon applies to different scopes of life. It definitely counts in the field of information and knowledge: The users of new information and communication systems are the same people who already used the old ones. New sources of information support primarily the people who are already familiar with using information. In 1978, the French report "The Computerization of Society" by Nora and Minc (1980) was given much attention. It warned that the information society will lead to conflicts between the information competent citizens and the digital illiterates. Inequality may even cause a revolution of the information poor against the information rich. Anti-intellectual movements, electronic book burning, cultural revolutions may occur. One way or the other, the new information and communication systems could finally bring an end of the leading role of science and knowledge in society.

3 Time of Oblivion

Documentation systems are means to bridge time and space in information and communication. Contrary to the spoken word, a written text can be read simultaneously in different places and at different times, unchanged over continents and centuries. The global distribution of texts is faster and cheaper in its electronic form than the paper based canals. However, the physical stability of the data storage medium of texts over time is not at all solved for electronic publishing, yet. This problem will hopefully be solved in the near future. There exist already now lots of methods for long-term data storage, but they are not in general use. We live in a time of transition. It is always the times of transition which leave stuff behind—any time of transition may become a time of oblivion. This was definitely the case when the era of the spoken language was replaced by the era of the written word. Another period of transition and therefore another period of oblivion have begun.

Changes from the primary information and communication system in society into another one do not need much time. Such a transition usually takes two generations, 50–60 years, to be completed. The basic process does not happen by learning within one generation, it only takes place when a new generation comes which grew up with the new information and communication environment. The generations up until the 1950s and 1960s were mainly socialised by the spoken language and by written texts on paper. There were no computer screens in the households, television was not omnipresent. Fairy tales, read by grandmother, the first newspaper, the exchange of books in school—these were landmarks in the intellectual development of these generations. The generation of the 1980s and 1990s is intermediary: Television was common in their childhood, but computers, mobile phones and the internet were not ubiquitous in their forming years. In elementary school there was no computer education, at university level the first laptops arrived in the lecture rooms. Nowadays, 60 years later, children are confronted with computer screens and mobile phones long before they can read and write, even before they can speak. They are the first digital natives.

A well-documented historical example for such a development is the transition from the period of the spoken word into the era of the written language in ancient Greece around 500 BC. Socrates (469–399 BC) was completely integrated in the world of the spoken word. He was, most probably, able to read and write, but as far as we know, he did not practise it. Instead he warned the citizens of Athens from using written texts: They will destroy memory and real wisdom, they will make us trust in false theories. We know Socrates's polemics due to his pupil Plato (428–348 BC) who explained them in the dialogue "Phaedrus" (Plato 1997). Plato was the intermediary of his time. He produced written texts, but in the form of the dialogue, taken from the era of the spoken word. Plato's pupil Aristotle (384–322 BC) was already completely integrated in the world of the written word. He was a native in the world of paper.

In his text "Phaedrus", Socrates foresees that the change of the main tool of communication in society from the spoken language to the paper based written text will lead to a loss in collective and individual memory, to a big oblivion. The main reason of this effect is simply carelessness. In fact, when the written language was introduced, lots of content of spoken tradition was forgotten. Before the written text was in common use, the human brain was the most important storage medium for individual and also for common knowledge. Collective wisdom was transferred from generation to generation in a formalised way; sometimes word by word. Fairy tales, epic texts, proverbs, magic formulas, songs were the media; history, medicine, law, art and of course religion the content. This oral tradition did not only describe reality, it constituted reality. Law was spoken in public trials, the ownership of land and its boarders were defined and reassured by collective inspection, religion was transferred in ritual actions and spoken texts, poems and songs were distributed by performance over time and space.

When the collective memory of society changed into written form, some of the spoken knowledge has been transformed into written texts, new texts were created (like in religion: the Jewish, Christian and Islamic believes are based on books), but a good deal of the oral knowledge got lost. Only a few thoughtful people realised that a huge cultural treasure was to be saved: The Grimm-brothers wrote down some of the German fairy tales, Bela Bartok collected traditional music from Hungary, Viktor Zack the folk songs of Styria. Contents which were not transferred into a written form vanished within few generations.

Today we are confronted with a similar situation. Books, photographs, scientific data etc. are transformed into a digital form while the original sources often get lost—even if we know that the new, digital forms are by far less durable than paper. Magnetic storage vanishes in time, CD Rom discs are designed to last for no more than 20–30 years. Even if the storage devices stay intact over time, will they still be readable? Will the hardware, the software, the operating systems be available after one or two generations?

In theory, the electronic, digital system of information and communication may be as durable as the oral or written forms. Long-time storage can be operated on a technical basis by migration, emulation and copying at regular intervals. The critical point is the phase of transition from one system to the other. The main problem is: Who applies the new technologies in practise? Who does the preservation work? Who cares about transferring written knowledge into electronic forms? Of course banks do, insurance companies, public notaries save their documents. Collective knowledge will survive through big institutions like Google or Wikipedia. But the written texts (and pictures) of small enterprises, households, dissertations and master thesis, to name a few, will be subject of oblivion. From the French Revolution thousands of pamphlets, leaflets, newspapers, letters, diaries still exist and tell us about the ideas and arguments behind. What will be left in some years from the 2011 "Jasmine revolution" in Tunisia? Where will all the Facebook, Twitter, and SMS messages, which made this social development possible, be in 100 years? (Rauch 2012).

The loss of information through technical weaknesses may be a source for oblivion. But oversupply of information may have the same effect. An information abundance which quickly fills the available storage media, which blocks the in-boxes, which confronts us with hundreds of e-mails after a few days, all that may result in an undifferentiated clearance of the computer memory, too, and a loss of (maybe) valuable information. What will happen if all our personal information, texts, financial statements, documents, photographs, music are stored in a computer cloud with practical unlimited storage capacity? Also this problem has a theoretical solution. A clever hierarchy of data, expiration dates, a frequent evaluation of the content of storage media, regular security updates, all this could prevent the information overkill. But again, who really applies these methods in daily life?

It is not the oversupply of information itself which causes the problem. In the future, intelligent software agents will help us. New forms of information retrieval and information design will overcome the present difficulties. It is the time of transition—the two generations living and working now—which causes the problem. We are still organizing our work with structures and mechanisms of the era of print and paper. But we are already using new electronic devices for writing scientific texts, new storage media, and new forms of networking tools which need a different behaviour and new methods not developed so far. In the meantime, there is the danger of uncontrolled oblivion.

Finally, the new electronic forms of communication and information need a complex and vulnerable system of infrastructure to be operational. Electric power supply, communication infrastructure, positioning systems etc. have to work in the background. Without reliable supply of electricity the complex system of new information and communication technology will immediately break down; without power supply no computers, no mobile phones, no railways and no cars will be operational any more. The access to money, the access to knowledge, the access to property will come to an end. Internal and external security is not guaranteed any more. The critical infrastructure of our society is a fragile system even under normal and peaceful conditions. In case of natural catastrophes, of war or terrorism it might easily collapse and endanger scientific work and results as well.

However, oblivion is not necessarily negative. A "creative oblivion" might, like creative destruction, clear the way to new insights, new theories, and new forms of thinking. There are already some ideas what that could look like. One is connected with Big Data.

4 Doom of Causality?

As mentioned before, communication media influence the way we think and interact. The era of enlightenment was closely connected with the common availability of books. It needed the book as tool for multiplying ideas. The movement was caused by the book since reading books increased the number of educated and critically thinking people. Democracy needs mass media like newspapers, radio or television to disseminate ideas; these media enable a form of emancipation and participation which in the long run will not accept autocracy. New information and communication systems are not only necessary tools for new intellectual developments; they are often the basic causes and driving forces. If we change the basic means of information and communication in society, we change the way of thinking and behaving.

Books and papers are a perfect instrument to distribute scientific ideas unchanged over time and space. On the other hand, the common use of written language made it necessary to organize scientific argumentation in a strictly linear way. In using oral language, we can always enrich words with gestures, mimic, tone and other non-verbal elements to add further dimensions to it; to emphasize, to qualify something, to express doubt or conviction. In the written language, non-verbal supplementation is not possible any more. The written text cannot react to the audience. The written language loses dialogic quality. Socrates criticized this fact in his dialogue "Phaedrus", when he said "Writings are silent; they cannot speak, answer questions, or come to their own defence. Accordingly, the legitimate sister of this is, in fact, dialectic; it is the living, breathing discourse of one who knows, of which the written word can only be called an image. The one who knows uses the art of dialectic rather than writing" (Plato 1997, pp. 275e–276a). Or in Socrates own words: "The dialectician chooses a proper soul and plants and sows within it discourse accompanied by knowledge—discourse capable of helping itself as well as the man who planted it, which is not barren but produces a seed from which more discourse grows in the character of others. Such discourse makes the seed forever immortal and renders the man who has it happy as any human being can be" (Plato 1997, pp. 276e–277a).

This is, by the way, also the very idea of "university". The inseparable connection of teaching and the development of knowledge is an institutional form which has its roots in the era of the spoken word. On the other hand, science performed by universities, research institutes and libraries—which exchange and discuss scientific ideas all over the world—is the result of the common usage of the written language. This was, according to Socrates, a very harmful development, since the written word is an instrument for reminding, not remembering. "Future generations will hear much without being properly taught, and will appear wise but not be so, making them difficult to get along with" (Plato 1997, pp. 274e–275b).

What Socrates predicted happened very quickly, but was not only negative at all. Science became an additive, cumulative discipline. In Isaac Newton's well known quote "If I have seen further it is by standing on ye shoulders of Giants" (Newton 1996, p. 143), the giants are in fact piles of written books accumulated, which would never be possible by spoken words. However, as soon as we transfer the spoken argument into its written form, we put it in a very strict corset. The written text forces an extremely rigid linear form of presenting ideas. As one character ads up to the next like the pearls on a necklace, the words, the ideas and the arguments have to follow the same strict linear pattern. Argumentation in a written text means putting ideas in a strict linear order. This is sometimes misunderstood to be causality. It is not a coincidence that Aristotle was one of the earliest scientists using written texts and one of the fathers of the concept of causality.

The structural form of the paper based written text turned out to be a very solid fundament for science. Ideas and arguments could be transferred unchanged through time and space, which was not possible in the world of the spoken language. Therefore not only science changed its structures from the spoken to the written word. Also the legal system, economy, even religions were deeply affected by this technological change of our main means of information and communication.

Now the fundamental system of information and communication is changing again. The written word is replaced by an electronic, multimedia form which is not any more connected to a person (the speaker) or a physical unit (the book) but which exists in a virtual sphere, the cyberspace. This does not only happen to scientific texts, it is also the case for all other forms of data: Transactions, communications, local positions, travel information, technical parameters, even our money and the legal basis of property end up and meet in an undefined "cloud" of information. This phenomenon is called Big Data and has important consequences for science, too.

Viktor Mayer-Schönberger and Kenneth Cukier describe in their book "Big Data" three important changes for science caused by Big Data (Mayer-Schönberger and Cukier 2013).

The first is the diminishing role of sampling as a scientific method. Random sampling is necessary when the collection of data is time consuming and expensive. This was the case in social sciences and economy but also in other fields of science and humanities in the nineteenth and twentieth centuries. In the 1960s and 1970s machine readable files of textual data were rare and highly valued sources for science. From small samples theories and models were developed, and proved. Sampling was a necessary short-cut because we did not have access to the full data sets. Nowadays the situation has changed completely. It is possible now to get data from the whole population at low cost and within reasonable time. Therefore we need no longer samples, sampling techniques, and extrapolations. We just look into the full set of data.

The second consequence of using Big Data in science is the decreasing value of exactitude. If we extrapolate from small samples to large populations we have to quantify as exactly as possible. Carefully selected data and highly developed methods were the virtue of the day. But if we can easily have a look into the whole picture, we can accept some noise in the data.

These two changes lead to a third shift which is the reason for the possible end of the importance of causality: Causality may be replaced by correlation. It was Nobel laureate Kahneman (2011) who pointed to the fact that forming quick causalities was an important strategy for the survival of mankind in dangerous environments. We got used to thinking in terms of causality, even if there is none, rather than accept correlations. We did this as survival strategy in the jungle and we do so in everyday life up to now. Big Data will help us to unmask lots of pseudo causalities, not only in esoteric environments but also in everyday life and in science. Correlations have a major advantage: They emerge from using the instruments of mathematics. Big Data shows us facts—not reasons. This is in most cases enough. "Causality won't be discarded, but it is being knocked off its pedestal as the primary fountain of meaning. Big data turbocharges non-causal analyses, often replacing causal investigations." (Mayer-Schönberger and Cukier 2013).

Causalities being a basic element of theories, some authors even proclaim the end of models and theory. "There is now a better way. Petabytes allow us to say: 'Correlation is enough'. We can stop looking for models" (Anderson 2008). Theories are means to extrapolate the world from a small basis of data. We do not need these crutches if we can see the whole world via Big Data. This idea caused an important discussion since it does not only undermine causality and hypothesis, finally it questions rationality itself.

5 Conclusions

Within two generations the information society has left the world of science fiction and doomsday literature and arrived in the middle of our lives. It already changed important industries like newspaper, tourism, transportation, trade, and will affect all other aspects of life within a few years. The information society causes a big change and destruction in all parts of our world.

The most important change of the information society will be the effect on the way we think and on the organisational forms of collective knowledge and its development, which is science. The basis of our present scientific work is physical accumulation of knowledge through printed texts and the concept of causality as driving forces of scientific argumentation. Both elements of science are now changing. With the loss of written texts as proof of facts and causality as method of discourse the concept of rationality itself is weakened.

The consequences of such a development might be dangerous. Rationality is much more than a scientific method. Rationality is an attitude. Rationality means to accept argumentation and discussion to be methods for solving problems better than force, or believe, or uncritical tradition. Rationality as the fundamental social agreement is the basis of enlightenment, of democracy, of our legal system, and human rights. Therefore the actual changes in science, caused by new information and communication technologies, could reach much further than just influencing academic work and careers. They might shake the foundations of our present society to the core.

References

Anderson C (2008) The end of theory: the data deluge makes the scientific method obsolete. In: Wired Magazine 16.07. Condé Nast Publishers, 23 June 2008

DiePresse.com (2014) Götze wird treffen: Aufregung um Twitter-Schwindel zum WM-Finale. In: Die Presse – online. 14.7.2014

Habermas J (1981) Theorie des kommunikativen Handelns. Suhrkamp, Frankfurt am Main

Kahneman D (2011) Thinking, fast and slow. Macmillan, New York

Matthew 25:29 (2011) The Holy Bible. English Standard Version. Crossway Bibles, Wheaton

Mayer-Schönberger V, Cukier K (2013) Big data – a revolution that will transform how we live, work and think. John Murray, London

McLuhan M (1964) Understanding media: the extension of man. Mentor, New York

Merton RK (1968) The Matthew effect in science—the reward and communication systems of science are considered. Science 159(3810):56–63

Newton I (1996) Letter to Robert Hooke, February 5th 1675/76. In: Newton I (ed) Eine biographie. Spektrum Akademischer Verlag, Heidelberg/Berlin/Oxford

Nora S, Minc A (1980) The computerization of society. MIT Press, Cambridge

Plato (1997) Phaedrus. In: Cooper JM, Hutchinson DS (eds) Plato: complete works (trans: Nehamas A, Woodruf P). Hackett Publishing, Indianapolis

Rauch W (2012) Angst vor dem Vergessen. In: Goltschnigg D (ed) Angst – Lähmender Stillstand und Motor des Fortschritts. Stauffenburg, Tübingen, pp 135–141

Deepening the Scope of the "Economic Model": Functionalities, Structures, Mechanisms and Institutions

Stefan P. Schleicher

I think that change is really occurring with the young people. My young students overwhelmingly don't understand how people could have believed in the old models. That is good.—Joseph Stiglitz (Institute for New Economic Thinking, 2015)

1 Introduction

"What went wrong with economics" asked The Economist (2009) in view of the unfolding economic crisis and illustrated the presented evidence by a book that melts away like a block of ice and carries the title "Modern Economic Theory". By echoing renowned voices of academia with similar concerns Carraro et al. (2014) arrive at the conclusion that "It took the deepest economic and financial crisis since the Great Depression to provoke an open debate amongst macroeconomists as to whether the 'economic model' taught in economics programs is adequate." Without wanting to pretend a full answer to these looming questions about the fading relevance of economics as it seems to be understood inside and outside of institutions ranging from top-rated universities to powerful central banks, we want to fathom at least some directions into which essential contents of the "economic model" could advance.

S.P. Schleicher (✉)
Wegener Center for Climate and Global Change, University of Graz, Brandhofgasse 5, 8010 Graz, Austria
e-mail: stefan.schleicher@uni-graz.at
url: http://stefan.schleicher.wifo.ac.at/

B. Bednar-Friedl, J. Kleinert (eds.), *Dynamic Approaches to Global Economic Challenges*, DOI 10.1007/978-3-319-23324-6_9

1.1 What Might Have Gone Wrong with Economics

Is economics as taught in the mainstream curricula at universities still a place for searching and finding solutions to the pressing problems of this century? This may be questioned if we look at a number of recent self-critical remarks of economists with a standing of Nobel Laureates.

With respect to our understanding of the behavior of individual consumers and companies McFadden (2006) summarized that "homo economicus, sovereign in tastes, steely-eyed and point-on in perception of risk, and relentless in maximization of happiness, is a rare species". With regards to the macro perspective of our economies Krugman (2009) in his Lectures at the London School of Economics argued that "Most of what we've done in macroeconomics for the past 30 or so years has turned out to be spectacularly useless at best, and positively harmful at worst". With respect to the success of market driven policies Stiglitz (2009) concluded that "The countries that followed the neo-liberal policies, which focused on market fundamentalism and the idea that markets worked on their own, by and large failed".

Meanwhile these warnings got updates that were underpinned by constructive guidelines. McFadden (2013) reiterated that economists need to handle their discipline differently and pleads for an overhaul of the microeconomic foundations of economics by opening to other disciplines as psychology, neuroscience and anthropology. Krugman (2015) reminded that the Eurozone is facing an economic crisis with a track record that is worse than during the 1930s and for many paradoxically he advocates slashing the new policy mindset of austerity in favor of seemingly old-time economics of stimulating demand. A similar outspoken advice is given by Stiglitz et al. (2015) to US policy makers by emphasizing that instead of a piecemeal policy change we must rewrite the rules of our economy.

1.2 Why Economics Is Struggling with the Very-Long Term

The warning voices about the failure of mainstream economics to provide policy advice to seemingly well-known problems as manifest in the ongoing economic crises get surprising support from the rather sobering contribution of economics to a new set of problems that require a time horizon that so far has not been in the scope of this discipline. These are the issues under the heading of climate change and related bio-physical limits of planet earth.

This weakness has become particular evident in Working Group III of the Intergovernmental Panel on Climate Change (IPCC 2014) in their evaluation of mitigation strategies that might be compatible with a 2 °C temperature increase: "Scenarios in which all countries of the world begin mitigation immediately, there is a single global carbon price, and all key technologies are available, have been used as a cost-effective benchmark for estimating macroeconomic mitigation costs [. . .]. Under these assumptions, mitigation scenarios that reach atmospheric concentrations of about 450 ppm CO_2eq by 2100 entail [...] an annualized reduction of consumption growth by 0.04–0.14 (median: 0.06) percentage points over the century relative to annualized consumption growth in the baseline that is between 1.6 % and 3 % per year". This was presented to the media in a statement that "2 degree mitigation will cost 0.06% of GDP growth", or "nothing" within the margin of error.

Robert Pindyck (2015) summarized the so-called Integrated Assessment models (IAMs) that are used for this kind of economic evaluations of climate policies as "... hav[ing] crucial flaws that make them close to useless as tools for policy analysis". He arrived at this conclusion by pointing out the arbitrariness of functional forms and parameter values, the rather vague knowledge about the relationship between the atmospheric CO_2 concentration and the resulting temperature increase and in the sequel the similarly uncertain impact on economic activity as measured by GDP.

1.3 How a Few Conceptual Extensions Could Make Economics More Useful

Without claiming to provide a comprehensive solution to the inherent problems of economics we put forward the proposition that the majority of deficiencies in this discipline results from two self-imposed restrictions. The first restriction refers to the limited scope in the perception of economic activities by focusing mainly on reproducible goods (including services) and a very few resources, as by production reproducible capital and human capital. The second restriction results from the interwoven relationships that describe economic structures and the coordinating mechanisms which operate on these structures by postulating market relationships that quite often turn out to be too simplistic or non-existing.

We propose therefore two conceptual extensions. The first extension opens up the scope of economic activity both by introducing the functionalities of well-being and an extended list of stocks and flows of resources. The second extension separates the description of economic structures from the mechanisms that operate on them, which may be market or non-market based. Furthermore we will demonstrate how these extensions can be made operational in the context of analyzing the transition of energy systems.

2 Enlarging the Scope of Economic Activity

Mainstream economics increasingly appears to be blinkered by its scope and vocabulary that reflect the GDP-based accounting framework. The limits of this conceptual construct when it is used for evaluating well-being and social progress got new attention with the Stiglitz–Sen–Fitoussi-Report (Stiglitz et al. 2009, 2010). At least three deficiencies can be identified: first, there is no distinction between good and bad economic activities; second, there is no reporting about the use of many sensitive resources as exhaustible or natural capital; third, there is no information about the distribution of income and products. We want to indicate in the sequel, how at least the first two deficiencies could be addressed and how this would also have a bearing for the third.

2.1 Responding to the Issue of Well-Being by Introducing the Concept of Functionalities

Among the many attempts to arrive at better metrics for welfare that transgress the limits of conventional indicators as gross domestic product or consumption the concept of functionalities, which is related to Amartya Sen's capability approach to welfare (e.g. Sen 1999), has a high potential for being made operational. For Sen, welfare can be described by a set of indicators for beings and doings, as adequate nourishment, housing and avoiding premature mortality.

We suggest extending this reasoning by attaching to every economic activity a functionality that indicates the welfare quality of such an activity. This can be made visible by some self-evident examples: The consumption of food is related to the functionality of nourishment; buildings are related to the functionality of shelter; transport activities provide access to persons and goods; numerous activities serve health, education, and cultural experiences. This, therefore, could be a first itemization of *functionalities relevant for well-being*:

- basic needs (housing, nutrition),
- personal services (education, health) and
- information and communication (access to persons, goods, culture).

These functionalities result from stocks, as buildings and the infrastructure of the internet, and from flows, as energy and human activities. The choice of technologies is a key to the composition of flows and stocks for achieving a specific functionality. Buildings e.g. can be made self-sufficient as to energy flows; people, e.g. can meet via video conferences without needing resource-intensive travelling. We will show later that this notion of functionalities can be made very specific and operational for some issues, as the design and transformation of energy systems.

This view on functionalities enables valuable insights for the evaluation of economic policies by checking the impact of particular policy measures on the relevant functionalities. Ultimately relevant is, e.g. the desired functionality of a comfortable living space, which means more than just observing the investment costs for a building, or the access to persons and goods, which does not always require a transport activity.

2.2 Responding to the Issue of Resource Use by Extending the Resource List

Economic activities, both related to production and consumption, have impacts on resources. Mainstream economics, however, is typically considering only a very few resources, above all capital stocks for buildings and machinery that are related to reproducible resources since they can be replenished by production processes. The mounting evidence of a by economic activities induced climate change and loss of species, and the increasing conflicts about the availability of water triggered alarms that many other important resources have been just neglected in our understanding of economic activities.

As a second extension of the current GDP-based accounting framework, we suggest therefore an *inclusive list of resources*, categorized e.g. as

- reproducible (goods and services)
- human (skills)
- natural (water, soil, air)
- material (energetic and non-energetic)
- social (trust, cooperation)

This extended list of resources is in particular relevant for the evaluation of long-term perspectives for economic development which obviously call for a high-efficiency use of all resources and for limiting the use of those resources that are crucial for the life-support systems of planet earth.

2.3 Looking for an Encompassing Modeling Framework

Why these extensions of the scope of economic reasoning are so essential becomes evident if we put them into an analytical structure and compare it with a conventional design. We suggest therefore what we coin an *encompassing modeling framework* which essentially supports modeling practices with deepened structural designs and is described by four types of relationships.

Functionalities F are generated by the stock of resources R and the flow of reproducible resources used for consumption c:

$$F = T_F(R,\ c) \tag{1a}$$

The total volume of the flow of reproducible resources q is partitioned between consumption c and investment i:

$$c = q - i \tag{1b}$$

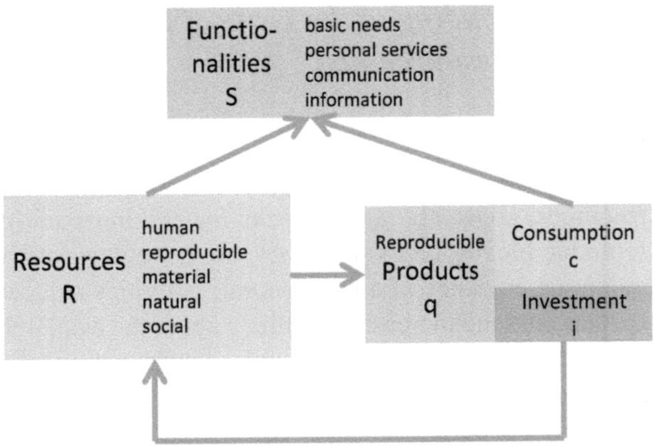

Fig. 1 Encompassing modeling framework. *Source*: Author

Reproducible resources q originate from the stock R and the flow r of resources from the inclusive list indicated above:

$$q = T_q(R,\ r) \tag{1c}$$

The stock of resources R is reduced by the amount of flows r needed for the production of reproducible resources and investment activities that also include regeneration of natural resources and recycling of materials and cleaning up natural resources $T_R(i)$.

$$R = R_{-1} - r + T_R(i) \tag{1d}$$

Remarkable for this encompassing modeling framework, as can be seen from Fig. 1, is the insight, that a certain level of well-being which is described by the set of functionalities F depends on the choices of three characteristic technologies: those that determine the amount of stocks and flows needed for a specific level of functionalities $T_F(.)$, those that produce similarly with stocks and flows the volume of reproducible resources which are synonymous with our conventional understanding of gross domestic product related flows $T_q(.)$, and those that are relevant for regenerating, recycling and reinvestment $T_R(.)$. These three categories of technologies finally determine the impact on all types of resources contained in the inclusive resource list (Fig. 1).

The merits of such an encompassing modeling framework (1) become evident when we compare it with a conventional modeling framework (2) which can be obtained just as a degenerate version of the encompassing model with three specific limitations.

First, emphasis is given to the flow of reproducible resources as measured by gross domestic product q, consumption c, and investment i. Well-being is closely tied to these flows:

$$c = q - i \tag{2a}$$

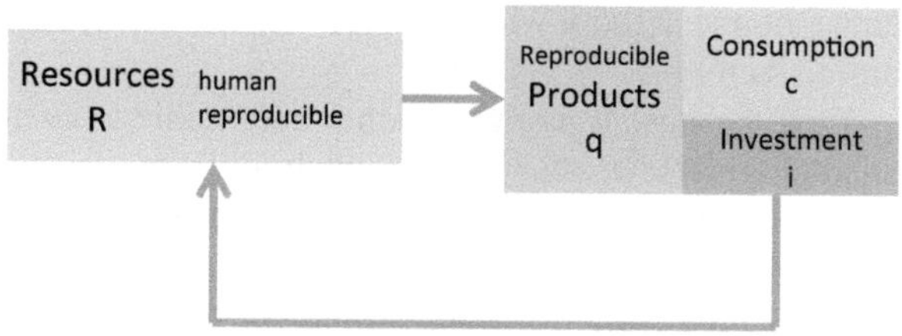

Fig. 2 Conventional modeling framework. *Source*: Author

Second, production of reproducible resources is mainly determined by the flows and stocks of reproducible and of human resources r_q, r_h, R_q and R_h, respectively:

$$q = T_q\left(r_q,\ r_h,\ R_q,\ R_h\right) \tag{2b}$$

Third, only the dynamics of the stock of reproducible resources with subtractions r_q and additions i during a production period are explicitly considered:

$$R_q = R_{q-1} - r_q + i \tag{2c}$$

The limited scope of the conventional model design becomes evident if we compare the corresponding Fig. 2 with the extended framework in Fig. 1. It turns out that it is this limited scope of the mainstream perspective that has caused many controversies about "green goals" and "green policies".

3 Separating Mechanisms from Structures

Up to now nothing in the modeling approach was said about coordinating mechanisms as markets and the related prices. This reflects a very deliberate separation in the modeling design between the representation of the physical structure of the system and the economic mechanisms that might operate on this structure.

The physical structure of our extended modeling approach encompasses the welfare relevant functionalities and the stocks and flows of resources that generate them with taking into account a comprehensive list of those resources. The resulting structure typically reflects effective mechanisms for coordination and incentives.

3.1 Structures May Reflect Complex Mechanisms for Coordination

Most conventional modeling approaches rely not only on market mechanisms but even intermingle in the model specification the representation of the structure of an economic activity with the market mechanism. For several reasons this is rather unsatisfactory.

First, because it might be rather impossible to evaluate in such a setting radical technological changes as, e.g. switching from conventional to almost energy-self-sufficient buildings. The complex interaction between the thermal structure of the building and the dependent energy flows for providing a particular thermal functionality requires a deepened structural specification and differentiated treatments of price impacts on investing and operating a building. Typically these multifaceted decisions are often summarized in a simple demand function for energy that depends on some price and income components.

There is in addition a second reason that recommends the separation from the specification of structures from the specification of mechanisms. This is related to the fact, that quite often market based mechanisms just don't exist. The choice of location of residential buildings may or may not reflect zoning regulations. The same holds for the thermal quality of those buildings.

3.2 Getting Prepared for New Institutional Settings

Another emerging evidence points at fundamental changes in the traditional role of consumers and companies. Consumers are discovering technologies that enable them to engage in production activities, e.g. by generating electricity from PV panels and selling the surplus electricity via the grid. Companies, on the other hand, are discovering the need to switch to new business models, e.g. by offering the services of a car instead of selling it as they were used to do.

New forms of cooperation and coordination of economic activities are becoming visible already in the most depressed states of Europe based on informal barter trade structures. Other institutional innovations concern the expansion of voluntary non-profit type organizations, in particular for taking care of the senior generation. Fairly established is already the institutional framework of sharing for some activities as services that compete with hotels and taxis.

All these institutional changes might have a major impact in the coming transformations of our economies and underline the need of modeling concepts that are able to separate structural changes from their institutional embedding.

4 A Case Study: The Transformation of the Energy System

The merits of this extended modeling design can be demonstrated for analyzing the transition of energy systems to high-efficiency and low-carbon structures in Austria (Köppl et al. 2014).

4.1 The Deepened Structure of an Energy System

By deepening the structural specification we discover a cascade four layers that constitute an energy system. The top layer deals with the energy related functionalities of the following types:

- thermal energy functionalities for maintaining buildings at comfortable temperatures and enabling heat-related production processes,
- mechanical energy functionalities for providing mobile or stationary services in all kinds of machinery, and
- specific electric energy functionalities needed for electric motors, lighting and electronics.

These energy functionalities are provided by useful energy which is characterized by its purpose as

- thermal use in low and high temperature processes,
- mechanical use in stationary and mobile applications, and
- specific electric use as in lighting and electronics.

The next layer of the energy system is composed of the energy flows that are metered in households and companies and which comprises final energy consumption for

- heating and cooling for buildings and production,
- fuels for stationary and mobile engines, and
- electricity for machinery, lighting, electronics and electro-chemical processes.

The amount of final energy is determined both by the amount of energy functionalities needed and the qualities of the corresponding application technologies as the thermal structure of buildings, the efficiency of machinery and appliances.

The lowest layer of the energy system concerns the primary energy flows as

- fossil energy (coal, crude oil, natural gas),
- renewable energies (thermal and PV solar, wind, hydro, biomass), and
- uranium for nuclear transformation processes.

4.2 The Technology Choices

Changing the existing structures of an energy system can be achieved by a wide spectrum of technology options which we classify according to their impact on the position of the energy cascade.

All changes in the layers of the cascade of the energy system are considered as technological changes. These technologies are defined in appropriate units and should be scalable. Some examples are:

- As to energy functionalities, the distances traveled by persons and goods in the mobility system; e.g. 1000 km of a person or a ton of goods.
- As to final energy consumption, the relevant application technologies, e.g. buildings with a specified thermal rating.
- As to primary energy, the transformation technologies used, e.g. a wind turbine with a specific rated generation capacity.

The technological and economic characteristics of a unit technology are described in a technology evaluation matrix. We deliberately differentiate between the investment and the operating phase for investigating the effects on flows, stocks, emissions and technology spillovers.

4.3 Developing Transition Strategies

Based on the toolbox of technologies and their economic impacts in a three-step procedure transition strategies can be developed:

- In a first step the long-term targets are specified, e.g. by 2050 the expansion of thermal functionalities for buildings by 20 % and a reduction target for greenhouse gas emissions of 80 %.
- In a second step trajectories to the present are specified based on assumptions about dissemination and learning curves.
- In a third step the economic impacts of these trajectories are traced, in particular their implications on investment requirements and the user costs for capital and operating.

4.4 How the Transition to a Low-Energy and Low-Carbon Energy System Looks Like

There are a number of insights which can be gained from this modelling approach which is based along the cascade structure of the energy system. Figure 3 summarizes the transition to low-energy and low-carbon structures based on such a deepened structural model of the sGAIN modeling family (Köppl and Schleicher 2014) for Austria. Total energy requirements for 2015 are scaled to 100. Thus currently 16 % of the inputs into the energy system are lost in transformation and distribution, 27 % are used for mobility, 22 % for low-temperature services in buildings and 17 % for high-temperature processes in production. Only 10 % of total energy supply is sufficient for all electric engines, for lighting and electronics. Finally 8 % of the energy requirements is needed for non-energetic use in industrial processes as iron and steel production.

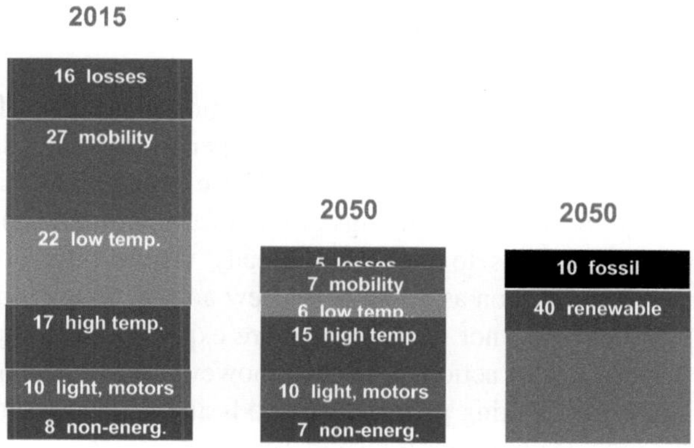

Fig. 3 Transition to low-energy and low-carbon structures for Austria. *Source*: Author

The model-based analysis for the transition to low-energy and low-carbon structures by 2050 reveals the high relevance of improving the energy productivity both in the application and the transformation technologies. This leads to an energy system that will provide substantially higher amounts of energy functionalities with less than half of the current energy flows. The main reductions in the flows can be harvested in the functionalities for low temperature, the building sector, and the functionalities for mobility, the transport sector. For final energy consumption electricity emerges as the leading type of energy. The primary energy mix shifts to renewables but in the case of Austria, which uses already 30 % of its energy supply from renewables, only an expansion of about one third of the current volume is needed in order to arrive at a reduction of CO_2 emissions of 80 % up to 2050.

The expected radical structural changes in our energy systems are much better supported by this type of modeling which follows the cascade structure of an energy system compared to conventional approaches that do not look into these layers of the energy cascade. The results obtained are fairly robust for economies with a similar economic structure than Austria. The economic impacts are compatible with long-run GDP growth rates in the range of 0.5–1.5 % p.a.

5 Some Tentative Conclusions for Policy Design

This extended scope of economic activities both with respect to the metric of welfare based on functionalities and a comprehensive view on resource use opens a number of insights for framing transition policies that can cope with break-through technologies and long-term targets as to limiting the use of sensitive resources.

5.1 Supporting the Design of Transition Strategies

First, such an extended view of interactions between welfare-relevant functionalities and an inclusive list of resources opens a better understanding about economic structures in the very-long-term. These insights concern, e.g. the way we want to design the stock of buildings or the infrastructure for mobility. The obvious answer for buildings is to make the already visible pilot projects for energy-sufficient buildings as soon as possible the new normal all the more, since these technologies turn out being not necessarily more expensive than those typically used. A much wider scope of actions is needed, however, for changing the infrastructure which is needed for linking persons and goods and which requires more than just switching the modal split of our transport system.

Second, policy strategies and related incentive mechanisms need to give more attention to the stocks, i.e. the infrastructure for buildings and mobility, than the flows, e.g. the consumption of energy. Even a high price of energy will not speed up sufficiently the switch to zero-energy or even plus-energy building designs. It is rather unconceivable that the big changes required in transport which span from zoning regulations to new urban designs and electric vehicles can be triggered just by a high carbon price. As a key message for policy design we realize the need for rebuilding and reinventing the infrastructure of our economies, ranging from the capital for buildings, transport and production to the human capital that will be needed to cope with these radical transformations. Policy actions will be needed ranging from targeted technology policies to innovative mechanisms for long-term financing (Aghion et al. 2009).

Third, the fundamental insight that we are able to provide in the very-long run the welfare-relevant functionalities of an economy with much lower flows of resources in general and without negative impacts on sensitive natural resources in particular requires decisive policy actions in the short-run that will stimulate for the foreseeable future the conventional economic indicators ranging from gross domestic product to employment.

5.2 Resolving Historical "Green" Controversies

A number of implications follow from the perception that the desired long-term structures for a resource-efficient and low-carbon economy require a major innovation effort visible both in rebuilding the quantity and the quality of existing capital stocks.

Above all a number of historic controversies in the context of environmental policies become redundant. This holds in particular for issues about the so-called double dividend—the potential beneficial impacts of policies both for the environment and conventional economic indicators—or the closely related issue of a trade-off between "green goals" and economic dynamics. These conflicts are resolved by realizing that plenty of opportunities are available for stimulating economic activity now in view of long-term economic structures that provide the functionalities for well-being with much lower resource use in the future.

The challenges for building an infrastructure that matches the targets of a resource-efficient and low-carbon economy are obvious. The next generation of buildings will integrate the functionalities of spaces for work and living but also provide the location for collecting and transforming energy. The current transport system will evolve to a mobility system which will require much less transport activities and will phase-out fossil fuels. Robotics and additive manufacturing, also known as 3D printing, will offer opportunities for a re-localization of production. Finally the energy system will experience a similar transformation as the transition from mainframe computing to personal computing some decades ago.

5.3 Setting New Agenda for Economic Policy

Summarizing the insights for new policy designs that emerge with an enriched "economic model", these could be in a nutshell the agenda for economic policies that are targeted to long-term viability and prosperity:

- *Re-measuring our economies.*
  Understanding the implications of economic policies could be substantially improved by extending the scope of the mainstream paradigm which is focused on GDP-related flows. These extensions concern a better understanding of economic activities both on well-being by introducing functionalities as the ultimate indicators and the footprint of economic actions by checking a comprehensive list of stocks which ranges from human, producible, to exhaustible and renewable materials but also to natural and social capital.
- *Re-discovering innovation.*
  The rather simple-minded approach of the mainstream paradigm to technological change needs to be challenged by welcoming the phenomenon of upcoming break-through-technologies and their implications in particular for those capital stocks which are considered as infrastructure for housing, mobility, production but also for improving human capital over the full life span. These innovations concern new business models, as access instead of ownership, but also societal qualities, as caring and sharing.
- *Re-writing institutions.*
  The still dominating market-driven paradigm is to be exposed to a fundamental reset by checking which institutional design is adequate in the sense of incentive compatibility for supporting the achievement of the desired functionalities and for harvesting innovation potentials. Obvious priorities are the role of financial institutions but also emerging cooperative non-market designs ranging from caring for children and seniors up to incubators for start-up companies.

In such an enriched economic paradigm transition policies will emerge that stimulate in the short-term economic growth as we are used to measure it but in the mid- and long-term will converge to a state with much lower flows, stocks of much higher productivity, and plenty of functionalities for satisficing well-being.

Acknowledgements I am deeply indebted to Karl Farmer with whom I have the privilege of being associated for more than three decades at the University of Graz. The considerations in this paper on new perspectives of economic reasoning reflect a journey that was decisively initiated by him and is still far from having arrived at a final destination.

References

Aghion P, Hemous D, Veugelers R (2009) No green growth without innovation. Bruegel Policy Brief, Issue 2009/07

Carraro C, Fay M, Galeotti M (2014) Greening economics: it is time. VOX, 26 April 2014

IPCC (2014) Summary for policymakers. In: Climate change 2014: mitigation of climate change. Contribution of working group III to the fifth assessment report of the intergovernmental panel on climate change. http://mitigation2014.org/report/summary-for-policy-makers

Köppl A, Schleicher S (2014) Energieperspektiven für Österreich. Zielorientierte Strukturen und Strategien. Austrian Institute of Economic Research, Vienna

Köppl A, Kettner C, Kletzan-Slamanig D, Schleicher S, Damm A, Steininger K, Wolkinger B, Schnitzer H, Titz M, Artner H, Karner A (2014) Energy transition in Austria: designing mitigation wedges. Energy Environ 25(2):281–304

Krugman P (2009) The return of depression economics. Lectures at the London School of Economics. https://www.youtube.com/watch?v=fKZKByt-wTk

Krugman P (2015) That Old-Time Economics. The New York Times, 17 April

McFadden D (2006) Free Markets and Fettered Consumers. AEA Presidential Address

McFadden D (2013) The new science of pleasure. NBER working paper no. 18687

Pindyck RS (2015) The use and misuse of models for climate policy. NBER Working Paper 21097

Sen A (1999) Commodities and capabilities. Oxford University Press, Oxford

Stiglitz JE (2009) Moving beyond market fundamentalism to a more balanced economy. Ann Public Coop Econ 80(3):345–360

Stiglitz JE, Fitoussi J-P, Sen A (2009) Report by the commission on the measurement of economic performance and social progress. http://www.stiglitz-sen-fitoussi.fr

Stiglitz JE, Fitoussi J-P, Sen A (2010) Mismeasuring our lives: why GDP doesn't add up. The New Press, New York

Stiglitz JE, Konzal M, Abernathy N, Hersh A, Kolmberg S (2015) Rewriting the rules of the American economy: an agenda for growth and shared prosperity. Roosevelt Institute, Washington

The Economist (2009) What went wrong with economics. 18 June 2009. http://www.economist.com/node/14031376

Is There a First-Mover Advantage in International Climate Policy?

Birgit Bednar-Friedl

1 Introduction

Climate change is a global environmental problem which affects people across the globe and across generations. As a consequence, mitigating climate change is an intertemporal global public good which encourages free-riding behavior by individuals, firms, and countries. While all are potential beneficiaries of less greenhouse gas emissions, every actor has an incentive not to contribute. But the more time passes by without serious efforts to address climate change, the higher will be the unavoidable damages triggered by climate change (GEA, 2012; IPCC, 2014).

The game-theoretic literature offers several explanations as to why so little has been achieved in international climate policy to date. One finding is that the global marginal benefits of abating an additional unit of emissions are much higher than the marginal benefits for an individual country. As a result, the non-cooperative (individually optimal) level of abatement will be much smaller than the cooperative (globally optimal) level (Barrett, 1994; Carraro and Siniscalco, 1998; Hoel, 1991). The higher the number of involved countries, the larger this difference between individual and collective marginal benefits is and hence also the difference between non-cooperative and cooperative abatement levels.

In symmetric games, i.e. when countries are identical both regarding potential impacts (marginal benefits of abatement) and regarding costs (marginal abatement costs), another key finding is that climate policy is best characterized as a prisoner's dilemma (Barrett, 2005; Finus, 2001). More recent literature, which is also supported by experimental evidence, however, shows that climate negotiations can turn

B. Bednar-Friedl (✉)
Department of Economics, University of Graz, Universitätsstraße 15, 8010 Graz, Austria
e-mail: birgit.friedl@uni-graz.at

B. Bednar-Friedl, J. Kleinert (eds.), *Dynamic Approaches to Global Economic Challenges*, DOI 10.1007/978-3-319-23324-6_10

into a coordination game: when the climate threshold, which when passed causes the climate system to change fundamentally, is known with certainty but the associated impacts are not, all countries have an incentive to coordinate; when in contrast the threshold is unknown, coordination fails (Barrett and Dannenberg, 2012).

But reality is more complicated than these public good games suggest. First of all, countries are non-symmetric. They differ in their abatement costs and hence economic efficiency arguments suggest that international emissions trading or other trading mechanisms of the Kyoto Protocol like the Clean Development Mechanism can ensure that a pre-defined global emission reduction target can be achieved at least cost. In a game-theoretic context, this implies that partial agreements, in which some regions form a climate agreement but others do not, become more likely (Asheim et al., 2006).

Another strand in the game-theoretic literature on climate policy investigates strategic behavior in instrument choice. For instance, Helm (2003) and Eyckmans and Kverndock (2010) investigate strategic behavior in international tradable permit markets, while Bárcena-Ruiz (2006) analyzes the advantage or disadvantage of sequential and simultaneous choices of emission taxes and MacKenzie (2011) and Schmidt and Marschinski (2010) those of emission permits.

Despite their valuable contributions, these game-theoretic models ignore linkages through international trade and hence results might be changed or even reversed when taking interactions on international commodity and asset markets into account (Carbone et al., 2009; Copeland and Taylor, 2005). A second shortcoming of the game-theoretic models is their partial equilibrium setup in which costs and benefits are directly linked to abatement levels (see, e.g., Buchner and Carraro, 2004; Eyckmans and Finus, 2007; Osmani and Tol, 2009; Pearce, 2003; Tol, 2001; Weikard et al., 2010). General equilibrium frameworks, in contrast, allow that abatement benefits and costs are based on economic fundamentals: technological differences such as emission intensities, dissimilar net foreign asset positions, and differences in household preferences with respect to global warming.

The aim of this contribution is to provide an overview of our own recent work in this field (Bednar-Friedl, 2012; Bednar-Friedl and Farmer, 2010, 2011). In particular, this chapter demonstrates how different positions to climate policy within the group of industrialized countries can be attributed to differences in net foreign asset positions as well as to environmental preferences. Moreover, we show how different positions of industrialized countries on the one side and developing countries on the other can be motivated by technological differences and different saving rates, in addition to differences in environmental preferences. Finally, we investigate the consequences of sequential policy setting.

This chapter is structured as follows: Before investigating the question of international burden sharing in a general equilibrium framework, we first look into the normative basis for burden sharing. Section 2 outlines the modeling framework. The consequences of simultaneous policy setting is analyzed for two industrialized countries in Sect. 3 and for two differently developed countries in Sect. 4. Sequential policy setting with the industrialized country as the first mover is the focus of Sect. 5.

The final section discusses the implications for real-world climate negotiations and concludes.

2 International Burden Sharing in Climate Policy

The question of international burden sharing has been at the center of climate negotiations since the beginnings of the United Nations Framework Convention on Climate Change (UNFCCC) in 1990. The idea of fair burden sharing is conceptualized in Article 3 of the UNFCCC. This article requires that countries should act in accordance with their “common but differentiated responsibilities and respective capabilities”. In the Kyoto Protocol (1997), this principle led to a differentiated approach between industrialized countries and developing countries. While industrialized countries were required to reduce their carbon emissions by 8 % on average by 2008–2012 relative to 1990 levels, developing countries were not subject to any reduction requirements.

With the Copenhagen Accord, adopted at the Conference of Parties (COP) 15 in 2009, the division into developing and developed countries started to fade. At the COP 17 in 2011, the Durban Platform emerged as a basis for a new international climate treaty for all countries. For the first time, it was agreed that not only developed countries should share the effort to achieve ambitious global emission reductions but that developing countries should be integrated as well, going considerably beyond the flexible mechanisms of the Kyoto Protocol.

On the way to designing a new international climate treaty to be adopted in Paris in 2015 and implemented by 2020, the burden sharing question therefore becomes increasingly important again. While it was already proposed in the early 1990s that emissions should globally contract and converge to equal per capita emissions (Meyer, 2010), the focus of the discussions then shifted to burden sharing among Annex I countries (e.g. den Elzen et al., 2010). With the recent developments after Durban, the focus has now broadened again to cover also middle and low income countries in burden sharing discussions.

Yet, little consensus has emerged on a common understanding of fair burden sharing. First, justice has both an intergenerational dimension, i.e., justice among different generations, as well as an intragenerational dimension, i.e., among differently developed countries. Second, justice can refer to procedures or to outcomes (consequentialist view), and can be seen as a question of distribution but also of compensation. Third, there is also little consensus on which burdens are to be shared: emissions reduction obligations, mitigation costs, adaptation costs, or welfare costs.

Table 1 summarizes a selection of principles and their operationalization as they are discussed in the normative literature (see, e.g., Kverndokk and Rose, 2008; Markandya, 2011).

Table 1 Fairness principles in international climate policy

Ethical concept	Principle	Description	Operationalization
Capability	Ability to pay	Are regions who are more able (i.e. richer) bearing higher burdens?	Different targets with reference to per capita income or wealth
Responsibility	Polluter pays	Are regions who pollute more bearing higher burdens?	Can refer either to current or historic emissions
Equality	Equal emissions	Are burdens distributed equally across regions?	Either equal reduction target (% reduction) or equal per capita target (carbon budget)

3 The Model

As this contribution summarizes the findings from three papers with different model versions (Bednar-Friedl, 2012; Bednar-Friedl and Farmer, 2010, 2011), we briefly describe the basic modeling approach which is a two-country two-good overlapping generations (OLG) model.[1] Each country is composed of perfectly competitive firms and finitely-lived consumers. Population in each country is constant and normalized to one.

To keep the model as simple as possible yet empirically relevant, both countries are perfectly symmetric in terms of endowments and consumption preferences but differ in production technology, environmental preferences, saving rates, and (per capita) public debt levels. We thus enable our model to differentiate within the group of industrialized countries (by diverging debt levels and environmental preferences) and between industrialized and emerging economies (by different production technologies, environmental preferences, and saving rates).

Each country produces a single good which is traded internationally. Per capita output in Home and Foreign are described by the following constant returns to scale technology:

$$x_t = M\,(k_t)^{\alpha_k}\,(p_t)^{\alpha_p}\,, \qquad\qquad y_t^* = M^*\,(k_t^*)^{\alpha_k^*}\,(p_t^*)^{\alpha_p^*}\,, \tag{1}$$

where M (M^*) denotes the productivity scalar in Home (Foreign), k_t (k_t^*) is the capital-labor ratio and p_t (p_t^*) is the pollution-labor ratio. When Home's production

[1]The model presented here is a combination and extension of Bednar-Friedl and Farmer (2010) and Bednar-Friedl and Farmer (2011). In Bednar-Friedl and Farmer (2010), production technologies differ across countries but preferences do not while in Bednar-Friedl and Farmer (2011) the reverse holds. In Bednar-Friedl (2012), which is the basis for Sects. 5 and 6, both production technologies and preferences differ across countries and saving rates are dissimilar. To keep the model still analytically tractable, we eliminate government bonds and assume instead that capital is internationally mobile.

elasticity of pollution exceeds Foreign's, $\alpha_p > \alpha_p^*$, then Home has a lower aggregate emission intensity.

Carbon emissions are regulated by national permit markets on which permits are auctioned. In each country, the long-lived government sets an exogenous cap on emissions: p in Home and p^* in Foreign. Profit maximizing firms therefore choose k_t^* and p_t^* to maximize their profits in each period:

$$\pi_t = x_t - q_t k_t - w_t + e_t p_t, \qquad \pi_t^* = x_t^* - q_t^* k_t^* - w_t^* + e_t^* p_t^* \tag{2}$$

Each country is inhabited by identical consumers who live for two periods, one working and one retirement period. The representative consumer's intertemporal utility depends on consumption during the working period, x_t^1 and y_t^1, consumption during the retirement period, x_{t+1}^2 and y_{t+1}^2, as well as on future global environmental quality E_{t+1}. E_{t+1} can be interpreted as the future climate which is affected by pollution in previous periods. For simplicity, the representative households in both countries have identical preferences across goods (denoted by the expenditure share $0 < \zeta < 1$), but different preferences over time (denoted by the rates of time preference $0 < \beta < 1$, $0 < \beta^* < 1$) and with regard to global environmental quality (denoted by the preference parameter $\xi > 0$, $\xi^* > 0$). Preferences can thus be represented by the following log-linear intertemporal utility function:

$$U_t = \zeta \ln x_t^1 + (1-\zeta) \ln y_t^1 + \beta \left[\zeta \ln x_{t+1}^2 + (1-\zeta) \ln y_{t+1}^2 + \xi \ln E_{t+1} \right], \tag{3}$$

$$U_t^* = \zeta \ln x_t^{*,1} + (1-\zeta) \ln y_t^{*,1} + \beta^* \left[\zeta \ln x_{t+1}^{*,2} + (1-\zeta) \ln y_{t+1}^{*,2} + \xi^* \ln E_{t+1} \right]. \tag{3*}$$

The domestic household thus maximizes intertemporal utility (3), while taking environmental quality as given. Following Jouvet et al. (2005), environmental quality is modeled as an international and intergenerational public good that is deteriorated by emissions caused by domestic and foreign firms:

$$E_{t+1} = \mu \bar{E} + (1-\mu) E_t - p_{t+1} - p_{t+1}^*, \tag{4}$$

where $0 < \mu < 1$ measures the speed of the autonomous pollution absorption and $\bar{E}$ is the pre-industrial state of the environment.

The budget constraint (in real and per-capita terms) of the domestic household when young is

$$x_t^1 + \frac{1}{h_t} y_t^1 + s_t = w_t - \tau_t, \quad s_t \equiv k_{t+1} + b_{t+1}^H + (1/h_t)\, b_{t+1}^{*,H}, \tag{5}$$

and when old is

$$x_{t+1}^2 + \frac{1}{h_{t+1}} y_{t+1}^2 = (1 + i_{t+1}) \left[k_{t+1} + b_{t+1}^H \right] + \left(1 + i_{t+1}^* \right) \frac{1}{h_{t+1}} b_{t+1}^{*,H}. \tag{6}$$

where τ_t is a lump-sum tax (or transfer) imposed by the national government, h_t denotes the terms of trade of Home (units of Foreign good per unit of Home's good), and $i_{t+1} = q_t - 1$ is the real interest rate. Household savings, denoted by s, are composed by the immobile domestic capital stock k_{t+1}^H, and by domestic and foreign bonds (b_{t+1}^H and $b_{t+1}^{*,H}$) which the household plans to hold at the beginning of its retirement period $t + 1$. Since government bonds are perfectly mobile across countries, the following no-arbitrage condition needs to hold:

$$\left(1 + i_{t+1}^*\right) \frac{h_t}{h_{t+1}} = \left(1 + i_{t+1}\right), \quad \forall t. \tag{7}$$

The corresponding budget constraints for the foreign consumer are:

$$h_t x_t^{*,1} + y_t^{*,1} + s_t^* = w_t^* - \tau_t^*, \quad s_t^* \equiv k_{t+1}^* + b_{t+1}^{*,F} + h_t b_{t+1}^F, \tag{5*}$$

$$h_{t+1} x_{t+1}^{*,2} + y_{t+1}^{*,2} = \left(1 + i_{t+1}^*\right)\left(k_{t+1}^* + b_{t+1}^{*,F}\right) + h_{t+1}\left(1 + i_{t+1}\right) b_{t+1}^F. \tag{6*}$$

As in Diamond (1965), we assume that the government runs a "constant-stock" fiscal policy and thus market clearing for Home and Foreign bonds demands:

$$b = b_t^H + b_t^F, \qquad b^* = b_t^{*,H} + b_t^{*,F}, \qquad \forall t. \tag{8}$$

To incorporate climate policy, we assume that the long-lived government sets a constant per capita emissions cap denoted by p in Home and by p^* in Foreign. The budget constraints for Home and Foreign governments require therefore that revenues from tax income and permit trading have to balance with interest payments to the bond holders:

$$\tau_t + e_t p = i_t b, \qquad \tau_t^* + e_t^* p^* = i_t^* b^*, \qquad \forall t. \tag{9}$$

To conclude the model, market clearing is required in each period for the domestic and foreign good, as well as for the world asset market:

$$s_t + \frac{1}{h_t} s_t^* = k_{t+1} + b + \frac{1}{h_t}\left[k_{t+1}^* + b^*\right], \quad \forall t. \tag{10}$$

This equation thus relates the terms of trade movements to capital accumulation and to the levels of domestic and foreign public debt. To see how the terms of trade are related to the two countries' net foreign asset position, we define Home's net foreign asset position by $\phi_{t+1} \equiv k_{t+1} + b - s_t$ and Foreign's by $\phi_{t+1} \equiv k_{t+1}^* + b^* - s_t^*$, and thus we can rearrange Eq. (10) to:

$$h_t = -\frac{\phi_{t+1}^*}{\phi_{t+1}}, \quad \forall t. \tag{10'}$$

Since Home's terms of trade need to be positive, i.e. $h_t > 0$, Foreign is a net debtor ($\phi^*_{t+1} > 0$) when Home is a net creditor ($\phi_{t+1} < 0$) and vice versa.

Solving the firm's and household's optimization problems and taking account of the three market-clearing conditions and the international interest parity condition yields a three dimensional system of motion for the terms of trade, and the domestic and foreign capital stocks. The fourth equation of motion is that for environmental quality which is obtained by rearranging Eq. (4) and substituting for the respective emission caps.

4 Climate Policy Coordination Among Industrialized Countries

Regarding differences within the group of industrialized countries, there is a growing consensus that countries differ in their perception and hence in the perceived benefits of climate change (Böhringer and Vogt, 2004; Brechin, 2010; Lorenzoni and Pidgeon, 2006; Reiner et al., 2006). Another common assumption in the game-theoretic literature is that countries differ in their cost of mitigation (see, e.g., Eyckmans and Finus, 2007; Osmani and Tol, 2009). We argue that this cost difference is determined by different net foreign asset positions.

We therefore assume that technologies are similar for both countries, i.e., $M = M^*, \alpha_k = \alpha^*_k, \alpha_p = \alpha^*_p$. Countries thus differ only in their debt level and their environmental preferences. For the sake of exposition, we furthermore let Home be a net foreign creditor (like the EU-15) and Foreign be a net foreign debtor country (like the USA).

We now turn to the question of how differences in net foreign asset positions and environmental preferences affect the optimal choice of emission caps in each of the two countries. For that, we assume that each government chooses its emission cap in order to maximize domestic steady-state welfare of the representative young household by taking the policy decision of the other country as given.

Home's government thus seeks to maximize its welfare function $W(k, h, p, p^*)$ resulting from the indirect lifetime utility function $U(x^1, y^1, x^2, y^2, E)$, evaluated at the steady state values for domestic and foreign capital stocks, the terms of trade, and environmental quality. The corresponding first order condition for Home is thus:

$$\frac{\mathrm{d}W}{\mathrm{d}p} = W_k \frac{\partial k}{\partial p} + W_h \frac{\partial p}{\partial p} + W_p. \tag{11}$$

The domestic welfare effects of a change in its emission cap p comprise therefore economic components, via the capital accumulation and the terms of trade channels, and an environmental component.

As shown in Bednar-Friedl and Farmer (2010), a stricter domestic emission cap ($p \downarrow$) leads to a terms-of-trade improvement ($h \uparrow$) and since moreover $W_h > 0$, welfare increases through the terms of trade channel. On the other hand, a stricter emission cap reduces domestic capital intensity ($k \downarrow$) but the sign of W_k depends on the net foreign asset position of the country under consideration. Hence, welfare either decreases or increases through the capital accumulation channel. Finally, when environmental preferences are not too strong (ξ small), then also $W_p > 0$. In general, the total welfare effect is thus either positive or negative, depending on the relative magnitude of these different channels. Yet, under the additional assumption that environmental preferences are not too strong and that the Golden Rule applies, the net welfare effect of a stricter domestic emissions cap is negative for a net foreign creditor country like the EU-15 (Bednar-Friedl and Farmer, 2010, Prop. 2).

Proceeding similarly for Foreign, where the governmental welfare function is represented by $W^*\ (k^*, h, p, p^*)$, gives Foreign's welfare effect:

$$\frac{\mathrm{d}W^*}{\mathrm{d}p^*} = W^*_{k^*}\frac{\partial k^*}{\partial p^*} + W^*_h\frac{\partial h}{\partial p^*} + W^*_{p^*}. \tag{12}$$

Comparing the own welfare effects of setting a stricter domestic emission cap thus reveals that the welfare effects are stronger for a net foreign debtor country (Foreign in our case) (for the formal proof, see Bednar-Friedl and Farmer, 2010, Prop. 2).

Focusing on the special case of the Golden rule, i.e., assuming that Home has no government bonds such that $b = 0$ and choosing $b^* > 0$ accordingly, and assuming equal expenditure shares for Home and Foreign goods ($\zeta = 1-\zeta$), gives for Home's and Foreign's reaction functions:

$$P^H(p^*) = \frac{(1+\beta)\alpha_P\zeta}{\beta\xi(1-\alpha_K) + (1+\beta)\alpha_P\zeta}\left[\mu\bar{E} - p^*\right], \tag{13}$$

$$P^F(p^*) = \mu\bar{E} - \left[1 + \frac{\beta\xi^*(1-\alpha_K)}{(1+\beta)\alpha_P(C+1-\zeta)}\right]p^*, \tag{14}$$

where $C = [\sigma(1-\alpha_K) - \alpha_K]^2 / \left[\alpha_K - \sigma^2(1-\alpha_K)^2\right] > 0$ and $\sigma \equiv \beta/(1+\beta)$ denoting the saving rate.

Inspecting the reaction functions in (13) and (14) reveals a negative slope and therefore the emission caps are strategic substitutes. Each country's best response to a stricter emission cap in the other country is thus a less strict cap. This result emerges because of the global public good property of emission caps—if on the other hand the international spillover effect was very small, also the case of strategic complementarity would be possible (see e.g., Bárcena-Ruiz, 2006). Note, however, that the slopes of the two reaction functions are different: when Home is a net foreign creditor (which corresponds to $b = 0$) and has sufficiently strong environmental preferences, Home's best response to a stricter emission cap is a comparatively smaller increase in its own emission cap than would be the best response by Foreign (the net foreign debtor country since $b^* < 0$).

Equalizing Eqs. (13) and (14) yields the following relationship between Nash levels of domestic and foreign emission caps $(p^N, p^{*,N})$[2]

$$\frac{p^N}{p^{*,N}} = \frac{\xi^*}{\xi} \frac{\zeta}{\zeta + C}. \tag{15}$$

If Home is a net foreign creditor country and has moreover environmental preferences which are at least as strong as Foreign, than Home will choose a stricter (i.e., lower) emission cap than Foreign: $p^N < p^{*,N}$ (Bednar-Friedl and Farmer, 2012, Prop. 1). If instead, Home has considerably lower environmental preferences than Foreign, then Home will choose a higher emission cap than Foreign. In that case, while the positive net foreign asset position mandates a stricter cap, lower environmental preferences tend to loosen the cap and if environmental preferences are sufficiently weak, Home will eventually set a laxer target than Foreign. Finally, equal emission caps ($p^N = p^{*,N}$) will only emerge as a special case, namely when lower environmental preferences in Home just counteract the difference in the net foreign asset positions.

Having investigated how different policy choices within the group of industrialized countries are triggered by differences in net foreign asset positions and environmental preferences, let us now turn to climate policy choice among industrialized and developing countries.

5 Climate Policy Coordination Between Industrialized Countries and Emerging Economies

Regarding differences between industrialized countries and emerging economies like China and India, technological differences are of key importance: The carbon intensity is higher in emerging than in industrialized countries because industrial production is more energy intensive than services and since emerging economies have a comparatively smaller share of services (World Bank, 2010). In addition, labor income shares are higher and capital income shares are lower in emerging economies (Durlauf and Johnson, 1995). There are also remarkable differences in propensity to save: While gross national saving is around 20 % of GDP in advanced economies, China has a saving rate of nearly 50 % (IMF, 2011; Ma and Yi, 2010). Finally, there is also growing empirical evidence that public concern over global warming and hence preferences for climate policy differ between industrialized

[2]Note that the Nash equilibrium permit levels are not optimal from a global social planner perspective since at the Nash equilibrium $dW/dp = dW^*/dp^* = 0$ while a global welfare maximum demands equal slopes of welfare indifference curves: $(dW/dp^*) / (dW/dp) = (dW^*/dp^*) / (dW^*/dp)$.

countries and emerging ones (Brechin, 2010; Lorenzoni and Pidgeon, 2006; Reiner et al., 2006).

We therefore assume that industrialized countries (Home) and emerging economies (Foreign) have different technologies (with a higher emission intensity in less advanced countries), different environmental preferences (which can be higher or lower depending on a country's vulnerability to climate change), and diverging saving rates (with higher saving rates in emerging and lower saving rates in industrialized economies). To keep the analysis still analytically tractable, from now on we neglect government debt and assume instead that households save in terms of internationally mobile capital. This alters the domestic consumer's budget constraints Eqs. (5) and (6) to:

$$x_t^1 + \frac{1}{h_t} y_t^1 + s_t = w_t - \tau_t, \quad s_t \equiv k_{t+1}^H + k_{t+1}^{*,H}/h_{t+1}, \tag{5'}$$

$$x_{t+1}^2 + \frac{1}{h_{t+1}} y_{t+1}^2 = (1 + i_{t+1})\, k_{t+1}^H + \left(1 + i_{t+1}^*\right) \frac{1}{h_{t+1}} k_{t+1}^{*,H}, \tag{6'}$$

and analogously for Foreign.

Since real capital is perfectly mobile across Home and Foreign, again a real interest parity condition holds between the two countries:

$$\left(1 + i_{t+1}^*\right) \frac{h_t}{h_{t+1}} = (1 + i_{t+1}), \quad \forall t. \tag{7'}$$

Moreover, government budget constraints change to:

$$\tau_t + e_t p = 0, \qquad \tau_t^* + e_t^* p^* = 0, \qquad \forall t. \tag{9'}$$

Instead of the bonds markets, domestic and foreign capital markets need to clear: $k_t = k_t^H + k_t^F, k_t^* = k_t^{*,H} + k_t^{*,F}$, and the international asset market equilibrium changes to:

$$s_t + \frac{s_t^*}{h_t} = k_{t+1} + \frac{k_{t+1}^*}{h_t}, \quad \forall t. \tag{10'}$$

Finally, the net foreign asset position of country is determined now by the differences in saving rates and capital production shares across countries: when domestic savings are insufficient to meet capital demand in production, capital inflows are needed and hence Home is a net foreign debtor country while Foreign is a net foreign creditor country. Algebraically, $\alpha_k/(1-\alpha_k) \gtrless \sigma \wedge \alpha_k^*/(1-\alpha_k^*) \lessgtr \sigma^*$ is equivalent to $\Phi \gtrless 0 \wedge \Phi^* \lessgtr 0$ (Bednar-Friedl, 2012, Prop. 2).

Proceeding as for the previous case of two groups of industrialized countries and assuming again that the Golden Rule holds, which now requires that α_k and α_k^* are chosen such that $i = 0$, yields the following reaction function for Home (for its derivation, see Bednar-Friedl, 2012):

$$P^H(p^*) = \frac{(1+\beta)\alpha_p\zeta}{\beta\xi(1-\alpha_k)+(1+\beta)\alpha_p\zeta}\left[\mu\bar{E}-\delta p^*\right]. \qquad (13')$$

Proceeding similarly for Foreign, gives Foreign's reaction function:

$$P^F(p^*) = \mu\bar{E} - \left[1 + \frac{\beta^*\xi^*(1-\alpha_k^*)}{(1+\beta^*)\alpha_p^*(1-\zeta)}\right]\delta p^*. \qquad (14')$$

Equalizing Eqs. (13′)–(14′) and requiring again equal preferences for domestic and foreign products, $\zeta = 1-\zeta = 0.5$, the relative stringency levels of Nash emission caps is given by:

$$\frac{p^N}{p^{*,N}} = \delta\frac{\alpha_p}{\alpha_p^*}\frac{(1-\alpha_k^*)}{(1-\alpha_k)}\frac{\sigma^*}{\sigma}\frac{\xi^*}{\xi}\frac{\zeta}{(1-\zeta)}. \qquad (15')$$

As shown in Bednar-Friedl (2012, Prop. 3), if $\xi\sigma \geq \xi^*\sigma^*$ holds, then Home will choose a stricter emissions cap than Foreign: $p^N < p^{*,N}$. If on the other hand $\alpha_p/(1-\alpha_k) > \alpha_p^*/(1-\alpha_k^*)$ and $\xi\sigma < \xi^*\sigma^*$, Home will choose a less strict emission cap than Foreign: $p^N > p^{*,N}$.

The first case in which the Home (i.e., the industrialized country) Nash permit level will be lower than the Foreign (i.e., developing country) Nash permit level emerges when Home's relative pollution production share ($\alpha_p/(1-\alpha_k)$) is lower than Foreign's, and when Home does not have a lower preference for the environment than Foreign ($\xi \geq \xi^*$), and when its saving rates is not smaller ($\sigma \geq \sigma^*$). If $\xi > \xi^*$, the difference in environmental preferences reinforces the differences in production shares.

When on the other hand Home has considerably lower environmental preferences than Foreign, then it might be optimal for Home to choose a higher (i.e., less strict) permit level than Foreign. Note moreover that equal burden sharing ($p^N = p^{*,N}$) may result only as a very specific parameter constellation of environmental and consumption preferences as well as factor productivities which ensure that the right hand side of Eq. (15′) is equal to one.

6 Unilateral Climate Policy and Sequential Policy

We now turn to the question of whether the industrialized economy would choose a stricter or a less strict emission cap if it would set its policy first, taking account of the optimal choice of the emerging economy.

To derive Home's Stackelberg permit level, we insert Foreign's reaction function, i.e., Eq. (14′) solved for p^*, into Home's welfare maximization problem:

$$\max_{\{p\}} W\left(k(p,p^*), h(p,p^*), p, p^*\right), \tag{16}$$

$$\text{subject to } p^* = P^{*,F}(p) = -\frac{\alpha_p^*(1+\beta^*)(1-\zeta)\left[\mu\bar{E}-p\right]}{\delta\left[(1-\alpha_k^*)\beta^*\xi^* + \alpha_p^*(1+\beta^*)(1-\zeta)\right]}.$$

Proceeding as for the simultaneous solution, gives:

$$\frac{\mathrm{d}W}{\mathrm{d}p^S} = W_k\left[\frac{\mathrm{d}k}{\mathrm{d}p} + \frac{\mathrm{d}k}{\mathrm{d}p^*}\frac{\partial p^*}{\partial p}\right] + W_h\left[\frac{\mathrm{d}h}{\mathrm{d}p} + \frac{\mathrm{d}h}{\mathrm{d}p^*}\frac{\partial p^*}{\partial p}\right] + W_p + \frac{\partial U}{\partial E}\frac{\partial E}{\partial p^*}. \tag{17}$$

Comparing Eq. (17) to Eq. (13′) reveals that now Home additionally considers Foreign's response to its emission cap: $\partial p^*/\partial p = (P^{*,F})'$. Setting Eq. (17) equal to zero gives Home's Stackelberg emission cap, and substituting this expression into Foreign's reaction function gives Foreign's Stackelberg emission cap.

Focusing again on the special case of the Golden rule, i.e., α_k and α_k^* are chosen such that $i = 0$, for the fraction between Stackelberg and Nash emission caps in Home we obtain (Bednar-Friedl, 2012, Prop. 5):

$$\frac{p^S}{p^N} = \frac{\beta\xi(1+\beta^*)\alpha_p^*(1-\alpha_k)(1-\zeta)+D}{\beta^*\xi^*(1+\beta)\alpha_p^*(1-\alpha_k)(1-\zeta)+D}, \quad D \equiv \left[\alpha_p(1+\beta)\zeta + \beta(1-\alpha_k)\xi\right]. \tag{18}$$

The Stackelberg emission cap for Home (p^S) is thus larger than its Nash emission cap (p^N) when $\beta\xi(1+\beta^*) > \beta^*\xi^*(1+\beta)$. But this is equivalent to the condition $\sigma\xi > \sigma^*\xi^*$. Thus, if Home has a higher saving rate and/or a stronger environmental preference than Foreign, then Home as a first mover will choose a less stringent emission target than in a simultaneous move game.

Note that whether Home as a first mover chooses a larger or a smaller emission cap than in the simultaneous move setting depends on its environmental preferences as well as on its saving rate relative to Foreign. This is in contrast to the Nash solution where the relative capital (and hence pollution) production share is decisive for the stringency of Home's welfare maximizing emission cap relative to Foreign.

What does this higher domestic emission cap imply for Foreign's emission cap? Acknowledging the slope of foreign's reaction function, it can be shown that it lies in the interval $(-1, 0)$ (Bednar-Friedl, 2012, Prop. 4). Thus, when $p^S > p^N$, Foreign will tighten its emission cap in a sequential game ($p^{*,S} < p^{*,N}$) but total emissions are nevertheless higher: $p^S + p^{*,S} > p^N + p^{*,N}$.

If on the other hand $\sigma\xi < \sigma^*\xi^*$, then Home will choose a stricter emissions cap in a sequential setting, Foreign will respond with less strict emission cap and overall emissions will be smaller compared to the simultaneous move game. If $\sigma\xi = \sigma^*\xi^*$, then there is no difference between the sequential and the simultaneous move game.

Having seen that Home will either choose a more stringent or less stringent emission cap as a first mover, and that hence total emissions will fall or rise in comparison to the simultaneous setting, we now turn to welfare effects.

We know from Sect. 5 that the net welfare effect of a stricter domestic emissions cap for Home is composed of a positive environmental effect, a negative capital accumulation effect and a positive terms-of-trade effect. However, domestic welfare is also affected by the corresponding increase in Foreign's emission cap. This spillover welfare effect can be decomposed into a negative environmental effect, a positive capital accumulation effect and a negative terms of trade effect for Home. While in general the net welfare effect for Home is unclear, we can show that the effect is unambiguously positive.

To see why a change from the Nash to the Stackelberg solution is always welfare improving, we need to acknowledge that Home as a Stackelberg leader chooses the welfare maximizing emission cap which lies on Foreign's best reaction function. Hence, by definition,

$$W(p^S, p^{*,S}) = \max_p W(p, P^{*,F}(p)) \geq W(p^N, p^{*,F}(p^N)) = W(p^N, p^{*,N}). \quad (19)$$

As a consequence, the industrialized country as a first mover will only deviate from the Nash solution if doing so is welfare superior.

Foreign, the emerging economy, experiences a welfare gain in a sequential setting regardless of differences in environmental preferences and saving rates. This finding can be corroborated by the curvature of Foreign's welfare indifference curve. According to Bednar-Friedl (2012, Prop. 6), it can be shown that $\mathrm{dMRS}^*/\mathrm{d}S^* < 0$ when $\xi\sigma > \xi^*\sigma^*$. But since in that case $p^S > p^N$ and moreover Foreign's reaction function is negatively sloped, $(P^{*,F})' < 0$), it follows that $W^*(p^S, p^{*,S}) > W^*(p^N, p^{*,N})$. When instead $\xi\sigma < \xi^*\sigma^*$, then $\mathrm{dMRS}^*/\mathrm{d}S^* > 0$ and $p^S < p^N$ and hence again $W^*(p^S, p^{*,S}) > W^*(p^N, p^{*,N})$. For $\xi\sigma = \xi^*\sigma^*$, $p^S = p^N$ and $W^*(p^S, p^{*,S}) = W^*(p^N, p^{*,N})$.

7 Conclusions

This chapter discussed how different positions in the climate policy arena can be explained by economic characteristics such as external balance, technological differences, saving rates, and environmental preferences. The review focused on three dimensions: different positions within the group of industrialized countries, different positions between industrialized countries and emerging economies, and the question of sequential policy setting along the lines of 'common but differentiated responsibilities' of the UNFCCC.

Among industrialized countries, welfare costs are found to be higher for net foreign debtor countries like the United States than for net foreign creditor countries like the European Union (EU-15). The reason for this higher welfare cost in net foreign debtor countries can be traced back to the higher opportunity cost of mitigation over time (Bednar-Friedl and Farmer, 2010). Whether a net foreign debtor country will however choose a stricter or less strict emissions cap as compared to the net foreign creditor country, depends on whether differences in environmental preferences are stronger or less strong than differences in external balances and on how important environmental quality is for welfare in general.

While concern for climate change has been rising during recent decades, mitigation efforts in most industrialized countries have been quite limited which could be regarded as an indication of the limited importance it has for today's welfare. With the threat of approaching critical climate thresholds, this view is likely to be challenged in the future.

For differently advanced countries, we first looked again into simultaneous policy choice. Assuming that the industrialized country has a lower pollution production share and sufficiently strong environmental preferences, it will choose a stricter emissions cap than the emerging economy. However, in a sequential move game, the industrialized country would set a less strict emissions cap. This could be some indication for the claim that some industrialized countries have so far only implemented a 'symbolic' policy. On the other hand, we found that the emerging economy will respond with a slightly stricter emissions cap in a sequential setting. Finally, both the industrialized country and the emerging economy experience a welfare gain in a sequential move game with Home as first mover. This welfare gain, however, does not imply that total emissions are always lower than in the simultaneous move game. Only when the industrialized country has stricter environmental preferences and a not too high savings rate, then total emission will be lower when the industrialized country moves first.

In reality, the question of sequential versus simultaneous policy choice is often a political fact rather than an explicit choice. It is strongly related to the question of fair burden sharing, not only across differently developed countries but also over time. How policy setting can and needs to evolve to achieve low stabilization targets by the middle and end century, remains a task for future research.

References

Asheim GB, Froyn CB, Hovi J, Menz FC (2006) Regional versus global cooperation for climate control. J Environ Econ Manag 51:93–109

Bárcena-Ruiz JC (2006) Environmental taxes and first mover advantages. Environ Resour Econ 35:19–39

Barrett S (1994) Self-enforcing international environmental agreements. Oxf Econ Pap 46:878–894

Barrett S (2005) The theory of international environmental agreements. In: Mäler KG, Vincent J (ed) Handbook of Environmental Economics, vol 3, Chap. 28. Elsevier, Amsterdam, pp 1457–1516

Barrett S, Dannenberg A (2012) Climate negotiations under scientific uncertainty. Proc Natl Acad Sci USA 109(43):17,372–17,376

Bednar-Friedl B (2012) Climate policy targets in emerging and industrialized economies: the influence of technological differences, environmental preferences and propensity to save. Empirica 39(2):191–215

Bednar-Friedl B, Farmer K (2010) External balance, dynamic efficiency, and the welfare effects of unilateral and multilateral permit policies in interdependent economies. Econ Model 27(5):980–990

Bednar-Friedl B, Farmer K (2011) Nationally and internationally optimal climate policies: External balances versus environmental preferences. CESifo Econ Stud 57(3):432–457

Bednar-Friedl B, Farmer K (2012) Public debt, terms of trade and welfare: the role of capital production shares, external balances, and dynamic (in)efficiency. Int Econ J 26(2):317–349

Böhringer C, Vogt C (2004) The dismantling of a breakthrough: the Kyoto Protocol as symbolic policy. Eur J Polit Econ 20:597–617

Brechin SR (2010) Comparative public opinion and knowledge on global climatic change and the Kyoto Protocol: The U.S. versus the world? Int J Sociol Soc Policy 12:1–26

Buchner B, Carraro C (2004) Emission trading regimes and incentives to participate in international climate agreements. Eur Environ 14:276–289

Carbone JC, Helm C, Rutherford TF (2009) The case for international emission trade in the absence of cooperative climate policy. J Environ Econ Manag 58:266–280

Carraro C, Siniscalco D (1998) International environmental agreements: incentives and political economy. Eur Econ Rev 42(3–5):561–572

Copeland BR, Taylor MS (2005) Free trade and global warming: a trade theory view of the Kyoto Protocol. J Environ Econ Manag 49:205–234

Diamond P (1965) National debt in a neoclassical growth model. Am Econ Rev 55:1126–1150

Durlauf S, Johnson P (1995) Multiple regimes and cross–country growth behavious. J Appl Econ10:365–384

den Elzen M, Höhne N, Hagemann M, van Vliet J, van Vuuren D (2010) Sharing developed countries' post-2012 greenhouse gas emission reductions based on comparable efforts. Mitig Adapt Strateg Glob Chang 15(5):433–465

Eyckmans J, Finus M (2007) Measures to enhance the success of global climate treaties. Int Environ Agreements Politics Law Econ 7:73–97

Eyckmans J, Kverndock S (2010) Moral concerns on tradeable pollution permits in international environmental agreements. Ecol Econ 69:1814–1823

Finus M (2001) Game theory and international environmental cooperation. Edward Elgar, Cheltenham

GEA (2012) Global Energy Assessment—Toward a Sustainable Future. The International Institute for Applied Systems Analysis, Laxenburg. Cambridge University Press, Cambridge/New York

Helm C (2003) International emissions trading with endogenous allowance choices. J Public Econ 87:2737–2747

Hoel M (1991) Global environmental problems: the effects of unilateral actions taken by one country. J Environ Econ Manag 20:55–70

IMF (2011) World economic outlook. Technical Report April 2011, International Monetary Fund, Washington D.C.

IPCC (2014) Summary for policymakers. In: Field CB, Barros VR, Dokken DJ, Mach KJ, Mastrandrea MD, Bilir TE, Chatterjee M, Ebi KL, Estrada YO, Genova RC, Girma B, Kissel ES, Levy AN, MacCracken S, Mastrandrea PR, White LL (eds) Climate change 2014: impacts, adaptation, and vulnerability. Part A: global and sectoral aspects. Contribution of Working Group II to the fifth Assessment Report of the Intergovernmental Panel on Climate Change. Cambridge University Press, Cambridge/New York, pp 1–32

Jouvet PA, Michel P, Rotillon G (2005) Optimal growth with pollution: how to use pollution permits? J Econ Dyn Control 29(9):1597–1609

Kverndokk S, Rose A (2008) Equity and justice in global warming policy. Int Rev Environ Resour Econ 2(2):135–176

Lorenzoni I, Pidgeon NF (2006) Public views on climate change: European and USA perspectives. Clim Chang 77:73–95

Ma G, Yi W (2010) Why is the chinese saving rate so high? World Econ 12:1–26
MacKenzie I (2011) Tradable permit allocations and sequential choice. Resour Energy Econ 33:268–278
Markandya A (2011) Equity and distributional implications of climate change. World Dev 39(6):1051–1060
Meyer A (2010) Contraction and convergence: the global solution to climate change. Green Books, Foxhole
Osmani D, Tol R (2009) Toward farsightedly stable international environmental agreements. J Public Econ Theory 11:455–492
Pearce D (2003) The social cost of carbon and its policy implications. Oxf Rev Econ Policy 19:362–384
Reiner DM, Curry TE, Figueiredo MAD, Herzog HJ, Ansoabehere SD, Itaoka K, Johnsson F, Odenberger M (2006) American exceptionalism? similarities and differences in national attitudes toward energy policy and global warming. Environ Sci Technol 40(7):2093–2098
Schmidt RC, Marschinski R (2010) Can China benefit from adopting a binding emission agreement? Energy Policy 38:3763–3770
Tol R (2001) Climate coalitions in an integrated assessment model. Comput Econ18:159–172
Weikard HP, Dellink R, van Ierland E (2010) Renegotiations in the greenhouse. Environ Resour Econ 45:573s596
World Bank (2010) World development report 2010: ssdevelopment and climate change. Technical Report. World Bank, Washington D.C.

Environmental Policy in an Open Economy: Refocusing Climate Policy to Address International Trade Spillovers

Karl W. Steininger and Thomas Schinko

1 The Current Challenge in Climate Policy

Ongoing climate negotiations have revealed a significant inconsistency between current trends in global GHG emissions and the "likely chance of holding the increase in global average temperature below 2 °C or 1.5 °C above pre-industrial levels" (UNFCCC 2011). Since 2000, global CO_2 emissions have increased at a relatively strong rate of 3 % per annum, presenting a formidable challenge for policy makers seeking to keep below the 2 °C target (Peters et al. 2013). Thus, in order for there to be a realistic chance of staying below 2 °C (GEA 2012; IPCC 2013, 2014), the current set of policy instruments needs to be further enriched and developed. This article focuses on one particular aspect in such development, i.e. the impact of the rising share of greenhouse gas emissions that are associated to the fact that countries are linked by international trade activity and how this relates to the impact of national climate policies.

Strong growth in global GHG emissions continues despite ongoing developments in global climate policy, and despite related policy efforts at the national level. In 1992, the initial United Nations Framework Convention on Climate Change

K.W. Steininger (✉)
Department of Economics, University of Graz, Universitätsstraße 15, 8010 Graz, Austria

Wegener Center for Climate and Global Change, University of Graz, Brandhofgasse 5, 8010 Graz, Austria
e-mail: karl.steininger@uni-graz.at

T. Schinko
Wegener Center for Climate and Global Change, University of Graz, Brandhofgasse 5, 8010 Graz, Austria

International Institute for Applied Systems Analysis (IIASA), Schlossplatz 1, 2361 Laxenburg, Austria
e-mail: thomas.schinko@uni-graz.at; schinko@iiasa.ac.at

B. Bednar-Friedl, J. Kleinert (eds.), *Dynamic Approaches to Global Economic Challenges*, DOI 10.1007/978-3-319-23324-6_11

(UNFCCC) established the relevant objectives and framework for global climate policy. A few years later, in 1997, the associated Kyoto Protocol called for the implementation of specific emission reduction targets for developed countries. The Kyoto Protocol will likely meet its objective of stabilizing territorial CO_2 emissions in the developed countries, but global emissions in 2012 were 60 % higher than the emissions in 1990. The major growth in global CO_2 emissions has been from emerging economies, where emissions in 2011 were 2.6 times higher than in 1990 (Peters et al. 2013).

About a quarter of global fossil fuel CO_2 emissions are emitted during the production of goods that are ultimately consumed outside their country of origin (Davis and Caldeira 2010). As a result, quite a large share of emissions in developing and emerging economies are due to the consumption and investment that takes place in industrialized countries. Kander et al. (2015) argue that while various regions in the world (notably the US) show some success in reducing their CO_2 emission *growth* rate, at the same time this has given rise to an increase in emissions in other regions of the world (recently, notably China) due to their net import demand.

The contrasting trends in emission growth in developed and developing countries is likely to continue unabated should strong economic growth in the developing countries persist, and should emission mitigation policy be promoted unilaterally in the developed countries (e.g. the EU 80 % reduction by 2050, as indicated in the roadmap to a low carbon economy and the energy roadmap 2050). Thus, the success of unilateral climate policies in lowering territorial emissions in regulated countries such as (member states of) the EU (e.g. in the form of the Kyoto Protocol, or the EU-ETS) has had little discernible impact on global CO_2 emissions, and hence on the overarching objective of keeping the temperature increase below 2 °C (Peters et al. 2013).

It is generally accepted that the efficiency with which society mitigates anthropogenic climate change could be enhanced by the introduction of a more broad-based and globally-harmonized climate policy (Reilly et al. 1999; Clarke et al. 2009). To date, achieving the ideal policy has remained politically infeasible. A further attempt is to be made in late 2015, in Paris. However, even if this proves successful, an agreement will still not become effective before 2020. Hence, there is a clear need for alternative policy designs, i.e. for mechanisms that will work under a fragmented climate policy regime, such as the one we have today (Aldy and Stavins 2012). Under today's setting, i.e. a regime where no single, globally-harmonized policy approach has either been agreed upon or implemented, a whole range of policy questions arise (Steininger et al. 2014). Probably, the most important one concerns carbon leakage. Broadly speaking, the term carbon leakage is used to describe a shift in carbon emissions from one region to another region. There are two main concepts of leakage used in the discussion of emissions embodied in trade: first, "policy-induced" or "strong" carbon leakage, and second, "consumption-induced" or "weak" carbon leakage (e.g. Peters and Hertwich 2008; Droege 2011). The first concept focuses on the environmental effectiveness of climate policy in a situation where the stringency level set in one country is higher than that set by the country's trading partners. In such a setting, "dirty" production may relocate to the regions exhibiting lower levels of climate stringency, and thus part of the success in domestic greenhouse gas emission reduction is offset by the increase in emissions abroad. The amount of the offset is what carbon leakage is designed to measure. Both environmental, as well as economic issues (e.g. loss of competitiveness), are matters of concern here.

The second concept of leakage, consumption-induced leakage, is related to the overall trend in the global division of labor which has led to ever more emissions becoming embodied in imports and exports. This is particularly true for exports from emerging to industrialized economies. This trend, however, is not necessarily triggered by climate policy measures. More importantly, it also reflects international differences in comparative advantage and the decrease in trade barriers in recent decades. Still, those countries (or country groups) that have implemented relatively stringent climate policies (e.g. the EU or other OECD countries) are characterized by a quite significant imbalance in their carbon embodied in foreign trade: Carbon embodied in imported goods far outweighs the carbon embodied in exported goods. For a country with a relatively small fraction of basic industry, such as Switzerland, the net balance of carbon, traded implicitly in the form of foreign trade, doubles the level of domestic emissions. For countries with significant energy-intensive industries, such as France or Austria, there is an increase of almost 50 % in greenhouse gas emissions when their net foreign trade balance in carbon emissions is accounted for (Peters and Hertwich 2008; Munoz and Steininger 2010).

The focus of the present paper is the challenge to climate policy arising from this growing share of trade-related emissions.

The structure of the present article is as follows. In Sect. 2, we start by taking a closer look at the concept of leakage. As the need to develop an adequate database is clearly paramount, Sect. 3 then deals with questions concerning the adequate measurement of emissions in a world interconnected by international trade. Section 4 provides a detailed account of how the current set of policy measures needs to be augmented in order to address the challenges identified. Finally, Sect. 5 offers a summary of conclusions drawn.

2 Climate Policy in an Open Economy: The Issue of Carbon Leakage

Developed countries argue that strong, unilateral emission reductions although costly, still only offer rather limited progress with respect to global emission reduction, due, in part, to the impact of carbon leakage. In this context, it is important to distinguish between two types of carbon leakage:

- Strong carbon leakage (policy-induced leakage) is said to occur when emissions increase in unregulated countries directly as a result of climate policy regulation in other countries (e.g. where the EU ETS raises business costs, this may lead to a local contraction in business activity and an expansion of external activity in an unregulated country)

- Weak carbon leakage (consumption-induced leakage) occurs where emissions increase in unregulated countries in response to increased consumption in regulated countries, and the associated change in the global division of labor (e.g. increased consumption of consumable products is facilitated by the adoption of highly specialized production in an unregulated country)

To date, most of the existing literature has addressed the question of strong carbon leakage. This has generally been found to be relatively small at today's carbon prices. Böhringer et al. (2012a) compare 12 global models. For a 20 % unilateral emission reduction of Annex 1 countries (excluding Russia) they find a leakage rate of between 5 and 18 %, with a median value of 12 %. Several recent studies suggest that weak carbon leakage is relatively large, amounting to up to a quarter of overall global emissions. The reductions of territorial emissions in many developed countries are partially, and in some cases completely, offset by the increase in emissions associated with the production of imported products (e.g. Munoz and Steininger 2010; Peters et al. 2011; Wiebe et al. 2012; Aichele and Felbermayr 2012). Such figures have now made it clear that weak carbon leakage and the associated consumption patterns need to become a major consideration in the design of effective and comprehensive global climate action strategies.

One effective means of assessing weak carbon leakage is to compare the reported emission inventories (territorial-based or production-based emissions), with consumption-based emission inventories: Consumption-based emission inventories allocate emissions to the country of consumption and can be estimated by taking the amount of territorial-based emissions and deducting the (net) emission transfers. Emission transfers are the emissions embodied in exported goods and services minus the foreign emissions from the production of imported goods and services. The change in the net emission transfers over time is a measure of weak carbon leakage, with strong carbon leakage being a subset of this (Peters et al. 2011).

In the case of the EU27, territorial emissions (1990–2010) decreased at a rate of 0.4 % per year, while consumption-based emissions increased at around 0.1 % per year (Peters et al. 2012). The difference in emission growth across the two concepts is much higher in the 1995–2005 period (a period in which neither the collapse of the Soviet Union in the early 1990s, nor the global financial crisis in 2008–2012, had any influence). For Austria, territorial emissions (1997–2004) increased at a rate of 2.4 % per year, while consumption-based emissions increased at a rate of 3.3 % per year (Munoz and Steininger 2010; APCC 2014). Similar trends are found in most developed countries. Peters et al. (2011), give a detailed account of the most important global trading partner countries and regions.

A critical issue with the current territorial perspective concerning GHG emissions is that it fails to capture the broader role of consumption and international trade in driving global emission trends, thereby missing many important policy areas related to carbon-intensive imports. The stabilization of territorial emissions in many developed countries is being countered by the increase in consumption-based emissions stemming from increased levels of consumption and the associated shift in the global division of labor to regions with less environmentally efficient production.

A second crucial channel through which domestic climate policy may influence emission levels abroad operates via its effectiveness in reducing global fossil demand. Should, as a result, world fossil prices decline, there may then be a rise in fossil fuel demand (and emissions) in uncapped world regions. This form of leakage channel is termed the energy market channel (Droege 2011; Böhringer et al. 2012a). Climate policy also implies a redistribution of income, and this is also likely to cause changes in global emissions (the income leakage channel), and induce development of (clean) technologies, that (might) spill-over to other countries (the technology spill-over channel) (Barker et al. 2007; Steininger et al. 2014).

It is thus becoming increasingly clear that there is need to develop additional policy instruments in order to augment and complete the current available mechanisms. Merely focusing on production-based policy instruments is no longer sufficient.

3 Measuring Carbon Emissions

The attribution of emissions to specific countries can be done in several different ways. This means that the very same emission activity may show up in the carbon balance of different actors (or countries), depending on which accounting principle one follows. Each accounting principle acknowledges only a subset of the emissions influenced by any single country (Lininger 2015). Indeed, if national policies are to factor in climate change, then one of the key issues to take into account is how each country's national policies affect global emission change in total.

3.1 Emissions Accounting and Indicators Available

Current negotiations within the UNFCCC rely on the standard UNFCCC allocation rule that attributes all emissions to that location (or country) where the greenhouse gas is physically emitted in the course of production (territorial concept; production-based carbon accounting). There are the following clear alternatives. One could instead attribute emissions to the final consumer (agent or country) of the good, during the production of which GHGs had been emitted (consumption-based carbon accounting).

In contrast to the above, one could also argue that it is the countries where fossil fuels are extracted, which should be held responsible and thus allocate the emissions to the extracting nations (extraction-based carbon accounting).

Moving one step further, it might sound reasonable to argue that it is not only the extracting countries that profit from fossil fuel sales but also those countries where any of the subsequent steps in the further processing of those fuels takes place and which thereby generate value added. Hence, under this view, and as a third alternative, those countries should be held responsible, and be allocated the emissions, to which value added accrues. Such value may derive, for example, from the mining and processing of fossil fuels, or from the generation of fuel-driven transport services, or from any other activity in a global supply chain which at some point involves the creation of emissions. After all, it is exactly these countries which profit from the new income. This would thus be an income-based carbon accounting system.

In contrast to production-based accounting (where emissions are allocated to that territory where they physically occur), in extraction-based accounting emissions are all allocated to the countries of origin of fossil fuels, while in income-based accounting, emissions are allocated across the whole supply chain up to final consumption (and across the respective territories where value added occurs), with emission shares being determined by the respective shares in value added.

3.2 The Information Content of These Indicators

For an emission indicator to serve as a practical basis for climate negotiations it is necessary that:

1. The emissions attributed to each actor (or country) truly reflect how the actor's (country's) actions contribute to global emissions, and
2. The sum of all emission accounts adds up to global emissions.

While all four carbon accounting principles comply with the second condition, each has shortcomings with respect to the first condition. The major shortcomings are:

1. The production-based accounting principle fails to reflect the level of discretion available to a country in choosing its import supplier (and corresponding carbon intensity). With respect to both consumption and investment demand by international suppliers, the choice of which particular goods and services are imported may be driven by the latter's greenhouse gas intensity. This thus provides each country with a certain level of flexibility and scope of action which is not at all reflected under production-based accounting. Under the latter regime, the greenhouse gas intensity of imports plays no role whatsoever.
2. The consumption-based principle does not reflect the level of discretion one has over one's own exports (and thus over carbon intensity). A similar criticism to that mentioned above, but with respect to exports, can be raised for consumption-based accounting. Under such an accounting regime all emissions arising in the production of goods later exported are not reflected at all, even though a nation has a clear handle on regulating and influencing the production (and greenhouse gas intensity) of its exports.
3. The extraction-based principle and the income-based principle do not adequately capture either the technology (and efficiency) employed in further processing steps, nor the consumption choices pertaining in the country of final use. The critique here is related to the problem that choices of demand and of technology

may not correspond to the attribution of emissions made according to the origin of the fossil fuels (extraction-based) or to the respective value added shares along the fossil fuel supply chain (income-based). Thus, where several options are available, such concepts may fail to reflect specific emission shares.

We should not be surprised about the difficulties of incorporating true emission levels in any single accounting system. In a world in which no country is self-sufficient, the actors determining both the carbon intensity at each step of the product life cycle, or whether the product is produced and consumed at all, are clearly likely to be located in different countries. If we bring it all down to just the question of whether it is the producer of a carbon intensive good or the consumer, to whom the emissions should be attributed, we clearly cannot allocate the full amount of emissions to both if we wish to fulfill the second of the above criteria. If we split emissions across the two, however, we run into another problem. For example, let's assume a country is producing a carbon intensive good. If this country is held responsible only for a fraction of the emissions that arise in the production of this good, given that it exports the good to other countries (and its emissions are fully dependent on its own production technology), it will have a much lower incentive to clean up its export production than to clean up products destined for domestic consumption. This clean-up disincentive can be said to originate from "light weighting of emissions".

Given the above mentioned shortcomings of the four main accounting principles, several combinations of these have been proposed in the literature. One of the most prominent among these is to hold both the producing and consuming country responsible, and assign an emission share to each (Lenzen et al. 2007). While there is no agreement on how to determine such a share (Steininger et al. 2014), for practical purposes it has been suggested that the respective shares in economic rent (or value added) be drawn upon to govern the division (Lenzen et al. 2007). However, this concept clearly involves "light weighting of emissions", and thus works badly if we are seeking to indicate the complete emission reduction potential.

3.3 Many More Factors Determining Global Emissions

The ways in which a country can influence global emissions are, however, much richer than those covered so far, i.e. they reach far beyond production technology and demand choice for home production, for exports and for imports. For a country engaging in climate policy which has not been 'globally-harmonized', the literature informs us of at least four indirect impact channels through which national policy may influence global emissions. Thus, in addition to any direct reduction of domestic emissions, we also often see impacts through (1) the competitiveness or relocation channel, i.e. climate policy can trigger relocation of industry to other countries and increase emissions there; (2) the energy market channel, i.e. effective climate policy reduces global fossil demand and thus world fossil prices, which then raises the quantity of fossil fuel demanded (and emissions) in uncapped world regions; (3) the income channel, i.e. the redistribution of income following climate policy and the associated changes in global emissions; (4) the technology spillover channel, i.e. the induced development of (clean) technologies, that (might) spillover to other countries.

The strength of production-based accounting lies in its ability to cover all direct emissions (and their reduction). However, hardly any indirect channels are captured by this method (where climate policy works other than with emission caps, the energy market channel also has a home market effect, this being the only indirect channel applicable under this accounting principle). The consumption-based accounting method misses out on the direct effect of changes in export emissions. While it captures the indirect channels (all those mentioned above), it only captures that fraction that is measurable in terms of own imports. Extraction-based accounting fully covers the energy market channel, but only a fraction of the direct and indirect channels, at best. Income-based accounting covers only a fraction of each of the direct and indirect channels, as it attributes not the full emission amount generated to the respective countries.

Once cultural and other relatively intangible factors are taken into account, many more channels emerge, and all these may have some influence on global emissions. Take the case of a TV soap opera or a celebrity, whose carbon-free (or carbon-intensive) lifestyle is copied in other countries (Roser and Tomlinson 2014). This multitude of less tangible channels may be subsumed under the heading of "cultural or celebrity" channel.

4 Enlarging the Range of Policy Instruments

The differences in growth rates between production-based (or territorial) and consumption-based emissions stifles the effectiveness and the efficiency of the UNFCCC, the Kyoto Protocol, and of unilateral climate policies. All such programs aim at reducing production-based emissions while simultaneously failing to take account of consumption-based emissions. A similar argument can be made with respect to the divergences arising between production-based and income-based or extraction-based emissions. There is thus a clear need for policy adjustment in order to augment present production-based mechanisms. In Europe, for example, current mechanisms include the Emission Trading System, energy taxes, and command and control instruments such as car emission standards, etc. These points are covered in the following three sections, focusing on extensions by consumption-oriented, extraction-oriented, and income-oriented instruments, respectively.

4.1 Extension by Consumption-Oriented Policy Instruments

Consumption-based emissions include a share of the territorial emissions, and thus there is some overlap between production-based and consumption-based policy instruments (cf. Dietz et al. 2009). It is thus important to define a consumption-based policy instrument. We take consumption-based policy instruments to be policy instruments that influence consumption patterns such that both fewer national and fewer global GHG emissions result (cf. Dietz et al. 2013). These instruments do not replace, but are complementary to, policy instruments that only focus on national emissions. Thus, the main area of omission lies in the lack of policy instruments which are sufficient for addressing an international trade environment; i.e. policies that influence the production and consumption of products in global supply chains.

In developing policy instruments we have to start from what we observe concerning current emissions across the full supply chains of products. Recent work (Andrew and Peters 2013) has split the distribution of global GHG emissions for household consumption into 129 different countries and regions, and shows the respective shares of direct household and indirect (embodied) domestic and foreign emissions. The smallest share of emissions is for the direct household emissions (such as those arising from private driving, or from heating with natural gas). On average, such direct emissions represent 15 % of the associated global total. Thus, 85 % of emissions occur in industries which are engaged in producing household consumption products. Globally, on average, 63 % of the emissions occur in industries located in the country of consumption, while 22 % (ranging from 2 to 73 % depending on country) occur in industries located in foreign countries. The foreign share is generally larger for smaller countries such as Austria (Peters and Hertwich 2008) and is growing rapidly over time (Peters et al. 2011).

An advantage of consumption-based emissions is that they allow for a detailed analysis of the international supply chain and of carbon regulation. The identification of UK supply chains reveals that emissions allocated to consumption are intricately linked to emissions that occur in production, and that the structure of the supply chain for different products can be complex and quite varied. For example, for UK meat consumption the largest share of emissions (84 %) occurs on cattle farms, while for clothing, 26 % of emissions occur in China in the electricity, textiles, and chemicals sectors. We can readily identify that for these two common categories of consumer goods, very different types of policy intervention may be needed. In addition to this, it is also important to consider policy instruments on the producer side that limit the emissions allocated to consumption.

4.1.1 Consumption-Based Policy Instruments to Complement Existing Instruments

There are no clear nomenclatures for policy instruments that focus on different agents. We define consumption-based policy instruments as those that influence consumption patterns in that they lead to reductions in both national and global GHG emissions. This thus serves to delineate them from policy instruments that primarily address national GHG emissions. Consumption-based policy instruments might be consumer-orientated, focusing on final demand (e.g., carbon labeling, information programs, product standards), or producer-orientated, focusing on industry (e.g. consumption-based targets, free allocation to exporters, border tax adjustments, etc.); see e.g. Tojo and Lindhqvist (2010). From a systems perspective, consumption and production are intricately linked. This means that consumer-orientated instruments influence producers, and producer-orientated instruments influence consumers. The terminology used here is designed to reflect the agents that are most directly influenced by the policy instrument.

Several policy instruments targeting consumer behavior already exist. Identifying good practice is highly complex. The focus always needs to be placed on the overall goal, i.e. the desire to achieve an absolute reduction in global emissions. In our case, this is with respect to the whole global supply chain (Weber et al. 2013) and has to take the heterogeneity of consumer demands into account. Instruments that need to be analyzed range from soft instruments, such as policies designed to raise awareness and alter individual behavior (micro level adjustments), to economic instruments designed to alter institutional arrangements and social practices (the meso level). This can include both top-down and bottom-up instruments designed to redirect incentives, promote drivers, or overcome constraints.

4.1.2 Consumer-Oriented Policy Instruments

Increasingly, it would appear that consumers are willing to think about more than their immediate well-being. They are also motivated in their purchasing decisions by non-economic or ethical values e.g. questions of global solidarity, human and animal rights, or by environmental considerations (see e.g. Brunner 2007 regarding dietary practices). Consumer markets have thus become increasingly political. This is a key point for marketers and policy makers aiming to mobilize consumers and influence consumption patterns. A shift is occurring as environmental problems move from being a state or governmental concern to one which is the responsibility of members of the civil society and those wishing to engage in political action e.g. through boycotting or "buycotting". This has already been acknowledged in the recently established and rapidly growing research field of political consumerism (sf. Boström et al. 2005).

However, lessons learnt from existing research, especially in the fields of mobility and housing, can serve as important input and inspiration for the development of consumer-oriented policy instruments. In the field of sustainable mobility, regulatory instruments (e.g. car-free days), and market-based instruments (such as taxes), are now widely established. Information-based instruments (e.g. labels) which aim to enable consumers to make informed choices are also being increasingly employed and seem to be effective in targeting or directing household consumption. Also, policy makers have started to recognize that consumer decisions are based on much more than purely rational arguments. Thus, instruments designed to induce behavioral change, e.g. via social marketing, are now seen to be significant in contributing to the growth of sustainable consumption practices (Rubik and Gossen 2011).

4.2 *Extension by Extraction-Orientation*

Extraction-based carbon emissions accounting has its own take on the issue of responsibility for global carbon emissions. While the production-based as well as the consumption-based approaches focus on the latter stages of the supply chain, the extraction-based approach assumes that it is the countries to whom value added accrues from the mining and selling of fossil fuels, which should be held responsible, and thus allocated the emissions. Hence, a production-based policy orientation has to be fundamentally reshaped if climate policies are to be based on the extraction-based emissions accounting approach.

4.2.1 The Rationale for Extraction-Oriented Policy Instruments

With the exception of afforestation in the form of REDD (reducing emissions from deforestation and forest degradation), and attempts to foster carbon sequestration in the form of carbon capture and storage (CCS), the main policies for curbing anthropogenic CO_2 emissions currently being discussed under the UNFCCC, are measures on the demand-side for fossil fuels. Here, goods and services may be regarded as either 'production-oriented' or 'consumption-oriented' elements in policy instruments. While a cap & trade scheme can be classified as a pure production-oriented policy instrument, since it directly effects producers in the country of production (territorial approach), taxes on carbon emissions can be applied as both production-oriented policy instruments and consumption-oriented policy instruments.

While consumption-based accounting and associated policy measures have already been frequently discussed in the UNFCCC process and in public discourse, extraction-based accounting and extraction-oriented policy instruments, which focus on the extraction and thus the production of fossil fuels, have so far received relatively little attention. Recently however, new discourse has emerged in international climate policy focusing on the supply of fossil fuels as original emission source and hence as a natural target for climate policy. In September 2013, the IPCC reported in its Fifth Assessment Report (IPCC 2013) that the world could only emit about a further 1000 $GtCO_2$ if temperatures were to stay below the critical 2 °C threshold (assuming a probability level higher than 66 %). According to Carbon Tracker (2013) this remaining 'budget' represents a relatively small part of the carbon embodied in the world's known fossil fuel reserves of 2860 $GtCO_2$ (IEA 2012). The report thus suggests that about two thirds of existing proven global fossil fuel reserves will have to be left in the ground. Along these lines, UN climate chief Christina Figueres told the coal industry at the World Climate and Coal summit in November 2013, which took place parallel to COP 19 in Warsaw, that much of the world's remaining reserves need to be left underground if the world is to escape the worst of climate change (UNFCCC 2013). This new focus on the supply of fossil fuels could also lead to a substantial reconsideration of the importance of fossil fuel supply in climate policies.

A broad body of theoretical and empirical literature dating back to Felder and Rutherford (1993) has focused on the analysis of sub-global production-oriented climate policy instruments, such as emissions trading schemes or carbon taxes, and on targeting CO_2 emissions in fossil fuel combustion. Many empirical, model-based approaches rely on a top-down computable general equilibrium (CGE) analysis (e.g. Babiker 2005; Balistreri and Rutherford 2012; Bednar-Friedl et al. 2012a, b; Böhringer et al. 2012a; Böhringer 2000; Burniaux and Martins 2000, 2011; Fæhn and Bruvoll 2009; Fischer and Fox 2007; Paltsev 2001). The conclusions that can be drawn from these analyses are that sub-global production-oriented climate policies can have a substantial direct impact on the international competitiveness of domestic energy and trade intensive industries, as well as an indirect impact on global energy prices via a reduced domestic demand for fossil fuels. Both effects may lead to a relocation of energy intensive production and to carbon leakage, i.e.an increase in consumption of fossil fuels, and hence in the generation of CO_2 emissions in countries outside the actual policy regions. Carbon leakage thus has the potential to jeopardize the environmental effectiveness of sub-global climate policies. The competitiveness effect and the associated energy market effect represent the two most important channels for carbon leakage.

It has been shown that border carbon adjustment (BCA) policies can be effective instruments in protecting industries from a loss in international competitiveness and hence in reducing carbon leakage via the competitiveness channel (Böhringer et al. 2012b; Branger and Quirion 2014; Kuik and Hofkes 2010). However, BCA policies cannot directly tackle the energy market leakage channel, which has been identified as being the strongest leakage channel (Böhringer et al. 2012a; Boeters and Bollen 2012). A reduced demand for fossil fuels in a region implementing climate policy reduces the global fuel price and hence increases the quantity of fossil fuel demanded in non-policy-implementing regions. The overall result is a decrease in global carbon emissions which is lower than that induced by the fall in domestic emissions in the policy implementing region. Hence, global environmental effectiveness of sub-global climate policies, even in combination with BCA measures, cannot be ensured by applying production-oriented climate policies. Massarrat (2008) points out that it is incorrect to assume that a reduction in fossil fuel demand will automatically lead to a reduction in fossil fuel supply. Ultimately, the outcome in terms of environmental effectiveness crucially depends on the supply-side reaction to sub-global production-oriented climate policies.

Hence, given the substantial carbon leakage effects associated with sub-global production-oriented climate policies such as the Kyoto Protocol or the EU-ETS, extraction-oriented policy instruments, i.e. policies aimed at reducing the supply of fossil fuels instead of the use of fossil fuels, could be a viable option for a more environmentally effective climate policy (Harstad 2012).

4.2.2 Discussion of Extraction-Oriented Policy Instruments and Their Assessment in the Literature

In the previous section we argued that a move from production-oriented to consumption-oriented policy instruments would require implementing policy instruments that influence both the production and consumption of products that have global supply chains. Likewise, a move to extraction-oriented policy instruments would again require taking a different view of the global supply chain of fossil fuels, goods and services, and of carbon emissions.

Instead of mitigating carbon emissions at the end of the pipe, namely, by capping the allowed amount of CO_2 emissions and hence the demand for fossil fuels, extraction-oriented climate policies strive to tackle CO_2 emissions at their very root, i.e. the extraction or supply of fossil fuels. Extraction-oriented policy instruments could either be implemented in the form of depletion quotas or depletion taxes aiming at a reduction in the levels of domestic fossil fuel extraction, or in the form of the provision of financial compensation to foreign fossil fuel producing regions so as to induce them to permanently forego the exploitation of their resources (such as the Yasuni ITT initiative which was proposed in 2007 by Ecuador's President Rafael Correa; IUCN 2012).

If a country group (e.g. the Annex I countries) implementing domestic climate policies were not connected to the rest of the world via international trade, a production-oriented cap on carbon emissions (effective via the demand for fossil fuels) would have the same effect on the level of global carbon emissions as an extraction-oriented cap of the same size on fossil fuel production. This is because in such a setting, regional extraction has to equal regional consumption of fossil fuels. Thus, unilateral production-oriented and extraction-oriented climate policies would have the same effect.[1] The same reasoning holds true when, in order to remove the potential for free riding, all countries participate in a climate agreement (Fæhn et al. 2013; Hoel 1994).

However, in reality, no comprehensive global climate treaty is in place and there is no single country or group of countries which is not connected to world markets, be it fossil fuel markets or goods markets. Thus, implementing climate policies in one world region alone will trigger feedback effects via global goods and energy markets. Furthermore, these feedback effects will be different for production-oriented and extraction-oriented climate policies.

Thus, if not implemented within a comprehensive global agreement, even sub-global extraction-oriented policy instruments may eventually be prone to

[1]This reasoning holds true only if combustion-based CO_2 emissions are considered and only if it is assumed that the combustion of fossil fuels generates the same amount of CO_2 emissions irrespective of sector, i.e., irrespective of how or where they are burnt. In reality, however, some sectors also emit considerable amounts of CO_2 emissions which are not directly related to fossil fuel inputs in production (Bednar-Friedl et al. 2012b). Moreover, different combustion processes also lead to different CO_2 emissions. For example, some industrial processes do not combust the full amount of fossil fuels but store a fraction of carbon in the resulting production (Lee 2008).

environmental ineffectiveness, similar in effect to the carbon leakage triggered by production-oriented policy instruments. As international fuel prices increase due to the domestic reduction of fossil fuel extraction, non-policy countries (especially if they are major fossil fuel exporting countries) may respond by increasing their fossil fuel production. This would again increase the global fossil fuel supply, rendering the initial extraction-oriented policy partially ineffective on the global scale (Fæhn et al. 2013). In turn, even the desired domestic emission reduction in the policy region could be counterbalanced by increasing imports of fossil fuels. Moreover, from an economic point of view, the further intensification of resource extraction in major fossil fuel exporting countries may eventually weaken the latter's economic growth and development prospects as a result of the so-called resource curse, or paradox of plenty (e.g. Auty 1993; Sachs and Warner 1995).

So far, extraction-oriented climate policies designed to limit the production of fossil fuels under partial compliance, have not been subjected to much analysis in the academic literature, certainly not with respect to leakage effects. It is only recently that the supply-side of fossil fuels has gained prominence, as seen in the debate on the "green paradox" initiated by Sinn (2008). Based on extensions of the standard model of intertemporal depletion of exhaustible resources (e.g. fossil fuel deposits), it has been pointed out in the literature (e.g. Gerlagh 2011; Hoel 2012; Sinn 2008; van der Ploeg and Withagen 2012) that demand-side climate policies, such as escalating carbon pricing and the introduction of backstop technologies, may signal the gradual decarbonization of the economy to resource owners, and hence a pending decline in their future sales potential. This in turn may then provide an incentive for fossil fuel suppliers to increase the extraction of fossil fuels, and hence promote greater carbon combustion over the short term. The conclusion from this literature is that effective climate policy measures must succeed in flattening the carbon supply paths on the international energy markets. Hence, authors in this field argue that future climate policies should rather focus on the supply-side of fossil fuels and carbon emissions instead of on the consumption and demand-side.

Another strand of literature focuses on efficient climate policy in the presence of free rider behavior in sub-global climate policy settings, and derives the optimal (i.e. cost effective) combination of production-oriented and extraction-oriented shares in a policy region in order to achieve a given climate policy target (Fæhn et al. 2013; Hoel 1994; Bohm 1993). In a case study for the Norwegian oil market, Faehn et al. (2013) find that in order to attain a cost-effective combination of the two types of policies, the majority of domestic emission reductions should be provided by extraction-oriented measures, i.e. a reduction of domestic oil extraction.

Further research in various dimensions is needed to place the causes and effects of extraction-oriented climate policies within a broader, more holistic perspective. For example, there is a need to analyze in more detail the market structure of global fossil fuel markets, especially the role of private companies and financial markets. In addition, there is a need to look at specific response options for different countries and regions and how these responses vary and interact, and to explain how and why specific actors may choose specific response options. Moreover, the intertemporal dimension of the problem with respect to the trajectory of depletion and acquisition of resource stocks, needs to be integrated into a macroeconomic framework in order to derive more detailed results for the required transition to a global low carbon economy. Furthermore, it would be an interesting research task to identify and assess the concrete policies needed to counter the potentially substantial leakage effects induced by extraction-oriented policy instruments. This might, for example, entail analysis similar to the BCA studies of carbon leakage triggered by production-oriented policy instruments. Finally, practical hurdles in implementing extraction-oriented policy instruments have to be discussed in more detail. For example, a fossil fuel-producing country like Norway may have several good reasons for combining extraction-oriented and production-oriented policy instruments when attempting to achieve its national carbon abatement targets. However, the incentives to implement extraction-oriented policies remain highly unclear for large fossil fuel producers such as the OPEC countries (currently facing no mitigation targets). Topics like compensation for income foregone from fossil fuel sales, or high subsidies for the exploration and extraction of fossil fuels, all have to be addressed from a global perspective.

4.3 Extension by Income-Orientation

All three of the previously discussed accounting methods—production-based, consumption-based, and extraction-based accounting—only reflect a single point in the value chain of fossil fuels. The problem is that the fossil fuels may have been extracted in a different place compared to where they have been burned to produce goods and services, and that the latter may then be consumed in some other place entirely (Davis et al. 2011). This is one result of the intensification of international trade (Peters et al. 2011) and of the strong geographical concentration of the extraction of fossil fuels (Smil 2003). Consequently, value added is generated in numerous instances along the global fossil fuel supply chain, and one might argue that it is the specific countries deriving value added from fossil fuel use at each link in the global supply chain, which should be held responsible for the corresponding share of carbon emissions.

An extension of carbon accounts by income orientation explicitly takes into consideration the complete supply chain of global carbon emissions and the related value added. Starting from production-based emissions accounting, Davis et al. (2011) set out to depict an internally consistent carbon account linked to the value added along the fossil fuel supply chain by (1) tracing back the production-based emissions to the point of extraction by using trade data, and (2) by using a multi-region input–output (MRIO) model to establish the forward link to consumption-based emissions. They find that 10.2 billion tonnes of CO_2, i.e. 37 % of global emissions, are from fossil fuels traded internationally and an additional 6.4 billion tonnes of CO_2, or 23 % of global emissions, are embodied in traded goods.

Understanding the distribution of value added in the global fossil fuel supply chain and the different interests of actors at different links in the supply chain may be important in facilitating a global effort to effectively reduce carbon emissions. A more detailed understanding of the fossil fuel supply chain is likely to be highly informative concerning the incidence of the economic burden associated with various climate policies, and is thus important when considering how to shape international climate policies.

Income-oriented policy instruments might thus be imposed at any point in the supply chain of fossil fuel-related emissions. Economic theory would suggest that the economic burden of any climate policy will eventually be distributed along the supply chain according to the related price elasticities of supply and demand, regardless of where exactly in the supply chain the policy is implemented (e.g. Fullerton and Metcalf 2002). However, this theoretical reasoning strongly depends on market structures and on the ability of different economic actors to pass on the policy (e.g. tax) burden.

In terms of the overall efficiency of a specific climate policy both the points of intervention, as well as the sectors and countries implementing the climate policy, may all be crucial determinants. As argued in the previous section on extraction-oriented policy instruments, transaction costs may be substantially reduced when climate policies are imposed at the very beginning of the fossil fuel supply chain, i.e. at the stage of fossil fuel extraction, since the number of fossil fuel extracting countries is much smaller than the number of producers burning fossil fuels, and also smaller than the number of consumers buying goods and services based on fossil fuel inputs (Davis et al. 2011). Moreover, as a result of the relatively strong market concentration in the extractive industries, a price imposed on carbon by the few extracting countries would lead to a spreading of the economic burden of the regulation across the whole supply chain (Mansur 2010).

Also, with respect to environmental effectiveness, imposing climate policies on the very few parties extracting fossil fuels would drastically reduce—but not fully eliminate—the opportunity for carbon leakage (according to Davis et al. 2011, fossil fuels extracted in China, the United States, the Middle East, Russia, Canada, Australia, India, and Norway account for 67 % of global CO_2 emissions). Under such a policy regime, producers of goods and services may still move from one fossil fuel supplier to another, but overall extraction remains geographically fixed.

5 Conclusions

Looking at the current divergence in global patterns of greenhouse gas emissions, i.e. at the discrepancy for many countries between the decline in emissions when measured on a territorial basis, and the increase in emissions when measured on a consumption-basis, raises several questions. The main concern is whether the international community—operating in a setting largely devoid of global climate policy harmonization—will ever be able to achieve the 2° global warming target, or even come anywhere close. We have analyzed the issue here in terms of carbon leakage, in both of its major manifestations: strong carbon leakage, denoting emission relocation abroad due to more stringent local climate policy, and weak carbon leakage, viewed as a process originating from the general development of international specialization.

In order to complement the existing set of policy instruments so as to cope more adequately with future developments, we need to devise measures which reflect the increasing role of international trade in global emissions. We have argued here that none of the existing individual emission accounting systems on its own is sufficient. Neither the production-based system (as currently followed under UNFCCC guidelines), nor the consumption-based, extraction-based, or income-based systems, are up to the job of generating the necessary information for effective climate policy in the current sub-global climate regime. With respect to the specific question of policy instruments, we identified the core issue as being the development of consumption-oriented, extraction-oriented, and income-oriented policy instruments. All these are needed in order to address the development of global greenhouse gases adequately.

Acknowledgements The authors thank Karl Farmer for the extraordinary research and personal atmosphere that he created and continuously nourished—that we as well as our colleagues at the Economics Department at Graz University so much enjoyed. This is much more than one could ever hope for to find at any research environment. Furthermore, the authors thank Birgit Bednar-Friedl, Moritz Kammerlander, Ines Omann and Glen Peters for their help on various concepts developed and discussed here. Funding for this research was granted by the Austrian Climate and Energy Fund (Austrian Climate Research Program, project INNOVATE).

References

Aichele R, Felbermayr G (2012) Kyoto and the carbon footprint of nations. J Environ Econ Manag 63:336–354

Aldy JE, Stavins RN (2012) Climate negotiators create an opportunity for scholars. Science 337:1043–1044

Andrew RM, Peters GP (2013) A Multi-region Input–Output Table based on the Global Trade Analysis Project Database (GTAP-MRIO). Econ Syst Res 25:99–121

Auty R (1993) Sustaining development in mineral economies: the resource curse thesis. Routledge, London

Babiker H (2005) Climate change policy, market structure, and carbon leakage. J Int Econ 65:421–445

Balistreri EJ, Rutherford TF (2012) Subglobal carbon policy and the competitive selection of heterogeneous firms. Energy Econ 34(Suppl 2):190–197

Barker T, Junankan S, Pollitt H, Summerton P (2007) Carbon leakage from unilateral environmental tax reforms in Europe, 1995–2005. Energy Policy 35:6281–6292

Bednar-Friedl B, Kulmer V, Schinko T (2012a) The effectiveness of anti-leakage policies in the European Union: results for Austria. Empirica 39(2):233–260. doi:10.1007/s10663-012-9186-7

Bednar-Friedl B, Schinko T, Steininger K (2012b) The relevance of process emissions for carbon leakage: a comparison of unilateral climate policy options with and without border carbon adjustment. Energy Econ 34(Suppl 2):168–180

Boeters S, Bollen J (2012) Fossil fuel supply, leakage and the effectiveness of border measures in climate policy. Energy Econ 34(Suppl 2):181–189

Bohm P (1993) Incomplete international cooperation to reduce CO2 emissions: alternative policies. J Environ Econ Manag 24(3):258–271

Böhringer C (2000) Cooling down hot air: a global CGE analysis of post-Kyoto carbon abatement strategies. Energy Policy 28:779–789

Böhringer C, Balistreri EJ, Rutherford TF (2012a) The role of border carbon adjustment in unilateral climate policy: overview of an Energy Modeling Forum study (EMF 29). Energy Econ 34(Suppl 2):97–110

Böhringer C, Bye B, Faehn T, Rosendahl KE (2012b) Alternative designs for tariffs on embodied carbon: a global cost-effectiveness analysis. Energy Econ 34(Suppl 2):143–153

Boström M, Föllesdal A, Klintman M, Micheletti M, Sörensen MP (2005) Studying Political Consumerism. Political Consumerism: its motivations, power, and conditions in the Nordic countries and elsewhere. Proceedings from the 2nd international seminar on political consumerism, pp 9–24. Copenhagen: Nordic Council of Ministers, 2005, p. 529, ISBN: 92-893-1129-0

Branger F, Quirion P (2014) Would border carbon adjustments prevent carbon leakage and heavy industry competitiveness losses? Insights from a meta-analysis of recent economic studies. Ecol Econ 99:29–39

Brunner KM (2007) Ernährungspraktiken und nachhaltige Entwicklung – eine Einführung. In: Brunner KM, Geyer S, Jelenko M, Weiss W, Astleithner F (eds) Ernährungsalltag im Wandel – Chancen für Nachhaltigkeit. Springer, Wien, pp 1–38

Burniaux J-M, Martins JO (2000) Carbon emission leakages: a general equilibrium view. OECD Economics Department Working Papers, 242, OECD publishing. doi:10.1787/41053555

Burniaux J-M, Martins JO (2011) Carbon leakage: a general equilibrium view. Econ Theory. doi:10.1007/s00199-010-0598-y

Carbon Tracker (2013) Unburnable Carbon 2013: wasted capital and stranded assets. Carbon Tracker Initiative in collaboration with Grantham Research Institute. http://www.carbontracker.org/wp-content/uploads/2014/09/Unburnable-Carbon-2-Web-Version.pdf

Clarke L, Edmonds J, Krey V, Richels R, Rose S, Tavoni M (2009) International climate policy architectures: overview of the EMF 22 international scenarios. Energy Econ 31(Suppl 2):S64–S81

APCC (Austrian Panel on Climate Change), Austrian Assessment Report (2014) Summary for policy makers. In: APCC (ed) Österreichischer Sachstandsbericht Klimawandel. Austrian Academy of Sciences, Vienna

Davis SJ, Caldeira K (2010) Consumption-based accounting of CO_2 emissions. Proc Natl Acad Sci U S A 107:5687–5692

Davis SJ, Peters GP, Caldeira K (2011) The supply chain of CO_2 emissions. Proc Natl Acad Sci 108:18554–18559

Dietz T, Gardiner TG, Gilligan J, Stern CP, Vandenbergh PM (2009) Household actions can provide a behavioural wedge to rapidly reduce U.S. carbon emissions. Proc Natl Acad Sci U S A 106(44):18452–18456

Dietz T, Stern CP, Weber UE (2013) Reducing carbon-based energy consumption through changes in household behaviour. Daedeus (J Am Acad Arts Sci) 142(1):78–89. doi:10.1162/DAED_a_00186, http://www.mitpressjournals.org/doi/abs/10.1162/DAED_a_00186#.U96HeLF6rgY

Droege S (2011) Using border measures to address carbon flows. Clim Pol 11:1191–1201

Fæhn T, Bruvoll A (2009) Richer and cleaner at others' expense? Resour Energy Econ 31:103–122

Fæhn T, Hagem C, Lindholt L, Mæland S, Rosendahl KE (2013) Climate policies in a fossil fuel producing country. Demand versus supply side policies. Statistics Norway Research Department Discussion Paper No. 747

Felder S, Rutherford T (1993) Unilateral CO_2 reductions and carbon leakage: the effect of international trade in oil and basic materials. J Environ Econ Manag 25:162–176

Fischer C, Fox AK (2007) Output-based allocation of emissions permits for mitigation tax and trade interactions. Land Econ 83(4):575–599

Fullerton D, Metcalf GE (2002) Tax incidence. NBER Working Paper Series No. 8829, National Bureau of Economic Research, Cambridge, MA
GEA (2012) Global energy assessment – toward a sustainable future. Cambridge University Press/International Institute for Applied Systems Analysis, Cambridge/Laxenburg
Gerlagh R (2011) Too much oil. CESifo Econ Stud 57:79–102
Harstad B (2012) Buy coal! A case for supply-side environmental policy. J Polit Econ 120(1):77–115
Hoel M (1994) Efficient climate policy in the presence of free riders. J Environ Econ Manag 27(3):259–274
Hoel M (2012) Carbon taxes and the green paradox. In: Hahn RW, Ulph A (eds) Climate change and commonsense: essays in honour of Tom Schelling. Oxford University Press, Oxford
IEA (2012) World energy outlook 2012. International Energy Agency, Paris
IPCC (2013) Climate change 2013: the physical science basis, working group I contribution to the fifth assessment report of the intergovernmental panel on climate change. Oxford University Press, Oxford
IPCC (2014) Climate change 2014: mitigation of climate change, working group III contribution to the fifth assessment report of the intergovernmental panel on climate change. Oxford University Press, Oxford
IUCN (2012) Beyond oil fields to fields of dreams. International Union for Conservation of Nature. http://www.iucn.org/knowledge/multimedia/audio/?10956/yasuniittinitiative
Kander A, Jiborn M, Moran DD, Wiedmann TO (2015) National greenhouse-gas accounting for effective climate policy on international trade. Nat Clim Chang. doi:10.1038/nclimate2555
Kuik O, Hofkes M (2010) Border adjustment for European emissions trading: competitiveness and carbon leakage. Energy Policy 38(4):1741–1748
Lee H-L (2008) The Combustion-based CO2 Emissions Data for GTAP Version 7 Data Base. GTAP Resource #1143. https://www.gtap.agecon.purdue.edu/resources/download/4470.pdf
Lenzen M, Murray J, Sack F, Wiedmann T (2007) Shared producer and consumer responsibility—theory and practice. Ecol Econ 61:27–42
Lininger C (2015) Consumption-based approaches in international climate policy. Springer International Publishing, Switzerland, Basel
Mansur ET (2010) Upstream versus downstream implementation of climate policy. NBER Working Paper Series No. 16116, National Bureau of Economic Research, Cambridge, MA
Massarrat M (2008) Eine neue Philosophie des Klimaschutzes. In: Altvater E, Brunnengraeber A (eds) Ablasshandel gegen den Klimawandel? Marktbasierte Instrumente in der globalen Klimapolitik und ihre Alternativen. Reader des Wissenschaftlichen Beirats von attac. VSA-Verlag, Hamburg
Munoz P, Steininger KW (2010) Austria's CO_2 responsibility and the carbon content of its international trade. Ecol Econ 69(10):2003–2019. doi:10.1016/j.ecolecon.2010.05.017
Paltsev SV (2001) The Kyoto Protocol: regional and sectoral contributions to the carbon leakage. Energy J 22(4):53–79
Peters GP, Hertwich EG (2008) CO2 embodied in international trade with implications for global climate policy. Environ Sci Tech 42:1401–1407
Peters GP, Minx JC, Weber CL, Edenhofer O (2011) Growth in emission transfers via international trade from 1990 to 2008. Proc Natl Acad Sci 108:8903–8908
Peters GP, Davis SJ, Andrew R (2012) A synthesis of carbon in international trade. Biogeosciences 9:3247–3276
Peters GP, Andrew RM, Boden T, Canadell JG, Ciais P, Le Quere C, Marland G, Raupach MR, Wilson C (2013) The challenge to keep global warming below 2°C. Nat Clim Chang 3:4–6
Reilly J, Prinn R, Harnisch J, Fitzmaurice J, Jacoby H, Kicklighter D, Melillo J, Stone P, Sokolov A, Wang C (1999) Multi-gas assessment of the Kyoto Protocol. Nature 401:549–555
Roser D, Tomlinson L (2014) Trade policies and climate change: border carbon adjustments as a tool for a just global climate regime. Ancilla Iuris (Special Issue on International Law and Ethics):222–244

Rubik F, Gossen MG (2011) Policy instruments for sustainable mobility. CORPUS Discussion Paper 2. Institute for Ecological Economy Research (IÖW), CORPUS – Enhancing the Connectivity between Research and Policy-making in Sustainable Consumption. http://www.scp-knowledge.eu/sites/default/files/knowledge/attachments/Rubik_Gossen-CORPUS%20WP%203%20Mobility%20Discussion%20Paper%20II.pdf

Sachs JD, Warner AM (1995) NBER Working Paper 5398: natural resource abundance and economic growth. Harvard University, Cambridge MA. http://www.cid.harvard.edu/ciddata/warner_files/natresf5.pdf

Sinn H-W (2008) Public policies against global warming: a supply side approach. Int Tax Public Financ 15:360–394

Smil V (2003) Energy at the crossroads: global perspectives and uncertainties. MIT Press, Cambridge

Steininger KW, Lininger C, Droege S, Roser D, Tomlinson L, Meyer L (2014) Justice and cost effectiveness of consumption-based versus production-based approaches in the case of unilateral climate policies. Glob Environ Chang 24:75–87. doi:10.1016/j.gloenvcha.2013.10.005

Tojo N, Lindhqvist T (2010) Product-oriented environmental interventions. Bundling of Policy Instruments. IIIEE Reports. International Institute for Industrial Environmental Economics (IIIEE), Lund University, Sweden. http://www.lunduniversity.lu.se/lup/publication/1702353

UNFCCC (2011) Establishment of an Ad Hoc Working Group on the Durban Platform for Enhanced Action. Decision 1/CP.17 by the Conference of the Parties (COP). http://unfccc.int/resource/docs/2011/cop17/eng/09a01.pdf#page=2

UNFCCC (2013) World Coal Association International Coal & Climate Summit Warsaw, 18 November 2013. Keynote address by Christiana Figueres, Executive Secretary United Nations Framework Convention on Climate Change. http://unfccc.int/files/press/statements/application/pdf/20131811_cop19_coalassociation.pdf

van der Ploeg F, Withagen C (2012) Is there really a green paradox? J Environ Econ Manag 64:342–363

Weber M, Driessen P, Runhaar H (2013) Evaluating environmental policy instruments mixes; a methodology illustrated by noise policy in the Netherlands. J Environ Plan Manag 57(9):1381–1397

Wiebe K, Lutz C, Bruckner M, Giljum S (2012) Calculating energy-related CO_2 emissions embodied in international trade using a global input–output model. Econ Syst Res 24(2):113–139

MIX
Papier aus verantwortungsvollen Quellen
Paper from responsible sources
FSC® C105338

If you have any concerns about our products,
you can contact us on
ProductSafety@springernature.com

In case Publisher is established outside the EU,
the EU authorized representative is:
Springer Nature Customer Service Center GmbH
Europaplatz 3, 69115 Heidelberg, Germany

Printed by Libri Plureos GmbH
in Hamburg, Germany